Basic Cell Biology for Dentists and Medical Students

Angelika Böttger • Jasmin Moneer

Basic Cell Biology for Dentists and Medical Students

Angelika Böttger
Department Biology II
Ludwig-Maximilians-University Munich
Planegg-Martinsried, Bayern, Germany

Jasmin Moneer
Bad Heilbrunn, Germany

ISBN 978-3-032-03635-3 ISBN 978-3-032-03636-0 (eBook)
https://doi.org/10.1007/978-3-032-03636-0

This Springer imprint is published by the registered company Springer Nature Switzerland AG
The registered company address is: Gewerbestrasse 11, 6330 Cham, Switzerland

Preface

This book is based on a lecture series delivered to students of dentistry at the Ludwig-Maximilians-University Munich, Germany, during the years 2006–2023. It is divided into five chapters, beginning with a basic description of cells and cell organelles. In this first chapter, we refer to the history of cell research and to the development of cytological, biochemical, and genetic methods for cell biological research. The second chapter provides an introduction into the biological chemistry of cells, covering the four classes of cellular macromolecules: nucleic acids, proteins, carbohydrates, and lipids. The third chapter deals with the nature of membranes, vesicle transport within the secretory system of the cell and sorting pathways for cellular components. Moreover, it covers communication of cells with each other and with the environment, thereby explaining some important signal transduction pathways of G-protein-coupled receptors. The fourth chapter is dedicated to the cytoskeleton, which ensures the shape and architecture of the cell and also its mobility. Special attention is paid to muscle contraction. Moreover, cilia and flagella as well as cell adhesion and the extracellular matrix are covered. Here, with dentistry students in mind, the processes of biomineralization and the pathways for producing dentin and enamel are considered. The fifth and final chapter deals with cellular homeostasis within an organism. Cell division, the cell cycle, and the basic elements of cell cycle control, including ways of signal transduction for growth factor signaling are described. Moreover, molecular mechanisms regulating cell death in animal cells (apoptosis) are discussed. The chapter ends with a brief description of viruses and how they use the cellular machinery for their replication.

Throughout this book, attention is paid to the history of cell research in order for students to see the long and stony road towards our current wisdom about the biochemical nature of the architecture and functionality of cells. Cells are small; molecules are even smaller. They are excluded from our unaided sensual perception. An awareness is needed for the fact that all insights into the described cellular processes are entirely dependent on the methods developed for researching them. These include microscopical techniques, but also biochemical and genetic approaches. Thus, all knowledge gathered so far is obviously limited. This should always be kept in mind when developing new approaches for medical treatments of disease. In light of the recent application of genetically interfering techniques for vaccinating humans, it is extremely important for medical professionals and students to understand how these techniques of genetic engineering work, how far our knowledge really goes, and what dangers may therefore result from such applications.

In this book, an overview of the most important aspects of cellular life is given for students in early semesters. For further details, there are excellent textbooks available. The books that have contributed most to preparing lectures and to assembling this book should be pointed out. These are consecutive editions of: "Molecular Biology of the Cell" by Alberts, Heald, Johnson, Morgan, Raff, Roberts, and Walter (W.W. Norton & Company, New York) and "Molecular Cell Biology" by Lodish, Berk, Kaiser, Krieger, and Bretscher (Freeman & Company). In addition, this book was shaped by our own long-standing and very broad experiences in experimental molecular cell research. Moreover, we have many colleagues and friends who supported this work by donating images of their own experimental research for our illustrations. We have mentioned them in the individual figures and want to express our greatest gratitude for their generosity.

At the end, our hope should be expressed that between the lines of this book, the reader can see the open space where things live, that we do not know. Doubtlessly, cell biology has contributed a lot of knowledge for developing medical treatments. However, it often seems that, in consequence, scientific and public opinion consider our bodies as mere molecular machines. Disease can then be fixed just by turning a knob here and inserting a screw there. Even genetic interference is considered a handle for ensuring human health. This is a misconception. The forces that run the molecular pathways of living beings are still hidden from us. Questions like "what is life?" or "what is the difference between humans and animals" are not answerable only on the basis of the knowledge of biologists, chemists, and physicists. For elucidation, natural sciences and the humanities should jointly search for wisdom. Similarly, medical doctors should look far beyond molecules, cells, and prescribed drugs when treating their patients.

Our colleague Alfred Gierer expressed the limits of our knowledge beautifully in an article he wrote in 2002. We are very grateful for his permission to use his words as a motto for this book.

> A cloud is condensed water, a snowflake is frozen water, H_2O; there are no mysteries about the molecules involved, and yet this fact is not enough to allow any of us to understand the form of clouds or the beauty of snowflakes. (Alfred Gierer, *Journal of Biosciences*, 2002)

Planegg-Martinsried, Bayern, Germany — Angelika Böttger
Bad Heilbrunn, Bayern, Germany — Jasmin Moneer
June 2025

Competing Interests The authors have no competing interests to declare that are relevant to the content of this manuscript.

Contents

1 General Structure of Cells, Short History of Cell Biological Research and Methods for Investigating Cells

A cloud is condensed water, a snowflake is frozen water, H_2O; there are no mysteries about the molecules involved, and yet this fact is not enough to allow any of us to understand the form of clouds or the beauty of snowflakes. (Alfred Gierer J. Biosci. 2002, Fig. 1.1 Snow, ice and water drops)

Fig. 1.1 Snow, ice, and water drops

A. Böttger, J. Moneer, *Basic Cell Biology for Dentists and Medical Students*, https://doi.org/10.1007/978-3-032-03636-0_1

What You Will Learn in This Chapter

In this introductory chapter, we recall the historic path that led to the discovery of cells and the establishment of the cell theory by the end of the nineteenth century. We then conduct a journey through the cell where we consider the structure and function of all animal cell organelles. Finally, we will describe older and modern methods of cytological, biochemical, and genetic research in order for the reader to understand the basis on which our current knowledge in cell and molecular biology is founded.

1.1 History and Overview

The name *cell* was coined in 1665 by Robert Hooke, who had achieved a 30 times magnification of a layer of cork, which, as we know now, represents dead tissue where only cell walls have remained. Nevertheless, the dead cells had left empty spaces behind, which made the impression of pores or small chambers, hence cells. Hooke described his observations: "I could exceedingly plainly perceive it to be all perforated and porous... these pores, or **cells**, ... were indeed the first microscopic pores I ever saw, and perhaps, that were ever seen, for I had not met with any Writer or Person, that had made any mention of them before this" (Hooke, 1665).

At about the same time, Antoni van Leeuwenkook used glass drops that he made himself, achieving a 300 times magnification, which allowed him to see bacteria, protozoans, and spermatozoa. He described his findings in about 300 letters, many of which were published in the Royal Society's *Philosophical Transactions,* the first one dating from 1673 (an account of these letters is found in (Folkes, 1723)). At the beginning of the nineteenth century, microscopic techniques improved, allowing a resolution of about 1 μm. The Scottish botanist Robert Brown not only discovered "Brownian motion" but also was the first to microscopically detect nuclei. He investigated fertilization in *Orchidaceae* (orchids) and *Asclepiadaceae* (milkweed) and noticed a prominent structure, which he saw as a kernel and called it "nucleus," a name that stuck, Fig. 1.2a.

Microscopes provided the solution to the major problem of cell research caused by the small size of cells. With our bare eye, we can only tell two points apart when they have a distance of about 100 μm, one tenth of a millimeter, but not smaller. This so-called optical resolution is principally limited for light microscopes by a diffraction limit. When objects are observed with visible light of wavelengths between 400 and 800 nm, two points have to be separated by a distance of half of the wavelength of the light beam in order for us to detect them as separated items. This means that we can only see two objects as distinct when they are 200–400 nm apart (Schermelleh et al., 2019). To increase the contrast in specimens of tissues and body fluids observed by light microscopy, certain techniques were applied, e.g., hematoxylin/eosin staining. The basic dye hematoxylin stains acidic structures like DNA and ribosomes blue, whereas the acidic dye eosin stains basic proteins red, e.g. in the cytoplasm. This provides the blue-red tissue sections that anatomists and students usually use for their studies. Another stain uses the dye Methylene Blue Azur, where basophilic structures stain blue. We show examples of this "Richardson staining" in Fig. 1.2.

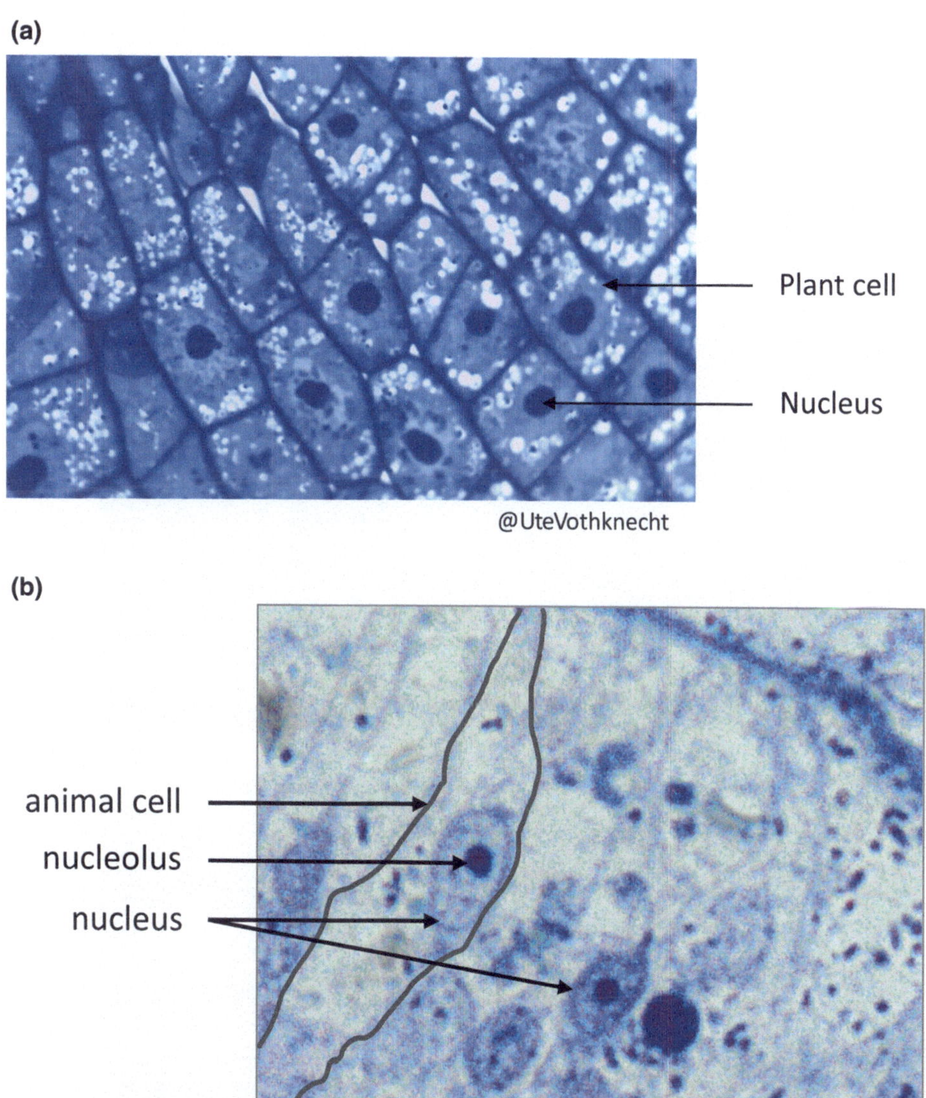

Fig. 1.2 Light microscopic images; (**a**) Section of plant root tissue (*Zea Mays*) after staining with 1% Azur in water and 1% methylene blue in 1% Bordex; (**b**) Semi-thin section of animal tissue (*Craspedacusta sowerbii*) after staining with Richardson staining showing epithelial cells and cell nuclei (light blue) with nucleoli (dark blue)

With light microscopy, especially after using histological staining procedures, we can easily recognize a cell nucleus with nucleoli inside an animal cell (Fig. 1.2b and size comparisons in Fig. 1.13). For further details, the resolution achieved by light microscopy is insufficient. It was possible to increase resolution by using smaller wavelengths, e.g., electron beams, in electron microscopy. Electron microscopy (EM) made it possible to

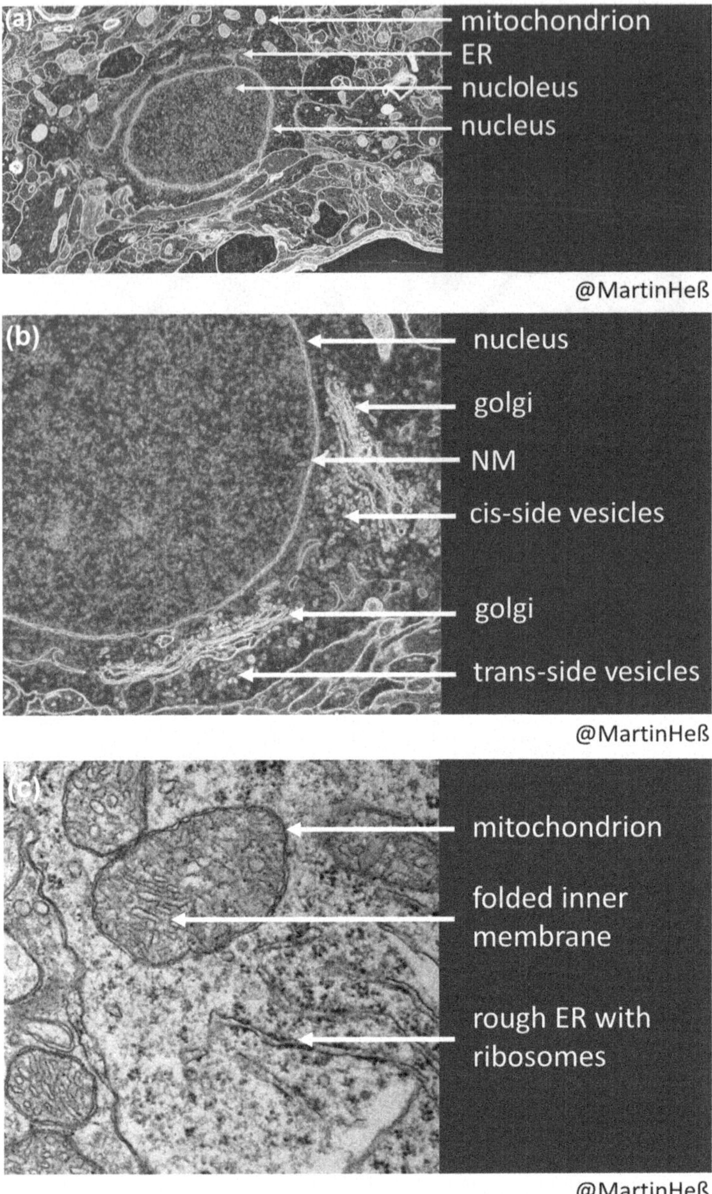

Fig. 1.3 Electron microscopic images; (**a**) Scanning EM of nucleus with nucleolus, endoplasmic reticulum (ER), and mitochondria; (**b**) Scanning EM of Golgi apparatus with vesicles on cis-side moving toward Golgi and on trans side, moving away from Golgi. Part of nucleus is also seen, NM nuclear membrane; (**c**) TransmissionEM, mitochondrion with smooth outer and folded inner membrane, rough ER membranes with ribosomes

image cell organelles including the ER, Golgi, and mitochondria. Small cellular structures like ribosomes (50 nm) and even single molecules, such as hemoglobin (diameter 5 nm), could now be recognized, Fig. 1.3a–c. In order to survive the harsh conditions in an electron microscope including the vacuum which has to be applied, samples have to be treated with chemicals which can change the structure of the cellular components that are analyzed. This poses a challenge for interpretation of the images, which can only be resolved by extended comparisons with data obtained by complementary methods. Therefore, light microscopy was improved too. By labeling objects with fluorescent dyes before microscopic analysis, cell organelles like mitochondria (Fig. 1.4c), endoplasmic reticulum, and the Golgi apparatus (Fig. 1.4a, b) become visible.

When cells were detected at 400–500 times magnification, they were soon considered as elementary structures of all tissues. This idea was first applied to plants and propagated by investigations of Matthias Jakob Schleiden and other researchers (Ribatti, 2018). In 1839, Theodor Schwann (1810–1882) characterized plant cells as follows: "Each cell is within certain boundaries an individuum, an independent entity. The live expressions of one cell are repeating themselves completely or partially in all others. These individual cells are, however, not mere aggregates but cooperate in a manner unknown to us in such a way that a harmonical whole body emerges", (*"Jede Zelle ist innerhalb einer gewissen Grenze ein Individuum, ein selbständiges Ganze. Die Lebenserscheinungen Einer Pflanzenzelle wiederholen sich ganz oder zum Theil in allen übrigen. Diese Individuen stehen aber nicht als ein bloßes Aggregat nebeneinander, sondern sie wirken auf eine uns unbekannte Weise in der Art zusammen, daß daraus ein harmonisches Ganze entsteht"),* (Schwann, 1939). He also described the major functional aspects of such cells including the formation of new cells, the extension (or growth) of existing cells, the metabolism of cellular content, and the secretion and resorption of cellular material.

Animal tissues seemed so different from plants and especially the occurrence of many diverse tissues nerved with blood vessels could originally not be associated with the idea of a basic building block, as which cells were considered in plants. By studying the *chorda cordalis* of frog larvae and fish gill cartilage, Theodor Schwann realized that these tissues were also built from cells, similar to plants. He saw the occurrence of a nucleus and a cell membrane as signatures of cells (Fig. 1.2). Like Schleiden, he thought that cells could arise from a liquid or unstructured substance where first a nucleus was formed and then the cell grew around it. He called this substance *Cytoblastema.* It was Robert Remak who doubted this way of cell creation. He observed cell divisions in blood cells of birds and eventually in embryos and remarked that "all embryonic cells multiply by division" (Remak, 1852). This idea was then formulated in a universal way by Rudolf Virchow in his *"omnis cellula e cellula,"* meaning that every cell originates from another cell (Virchow, 1858). This had implications for the hypothesis of the evolutionary history of life. When no cell is "de novo" created, we have to assume a universal ancestor cell from which all life is derived. This is indicated in the evolutionary tree of life in Fig. 1.5.

The evolutionary oldest cells we know are bacteria and *Archaea*; both are prokaryotes that lack membrane-surrounded cell organelles including a nucleus. Both come in

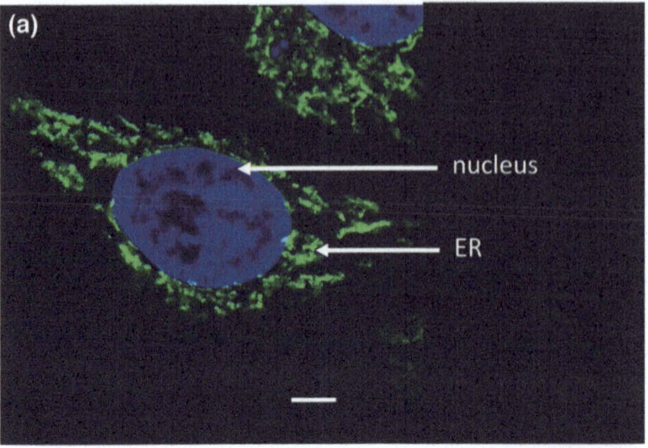

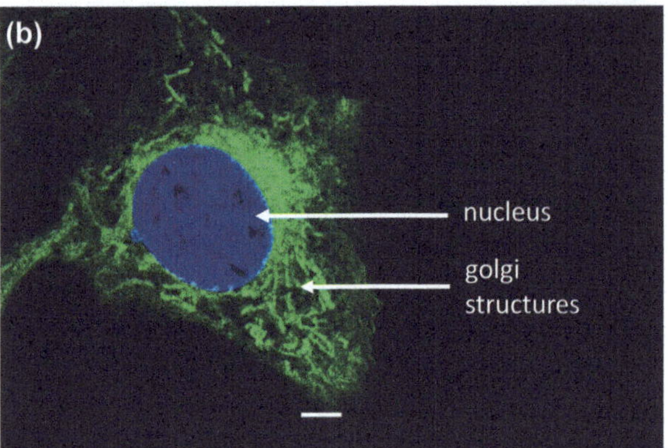

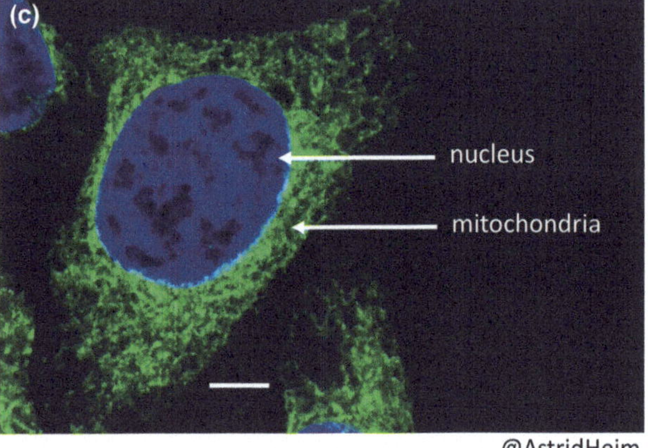

Fig. 1.4 Fluorescent microscopic images obtained by laser confocal microscopy; (**a**) Endoplasmic reticulum stained with anti-Calnexin antibody; (**b**) Golgi structures stained with Ceramid dye; (**c**) Mitochondria stained with anti-ATPase antibody; all scale bars 5 nm

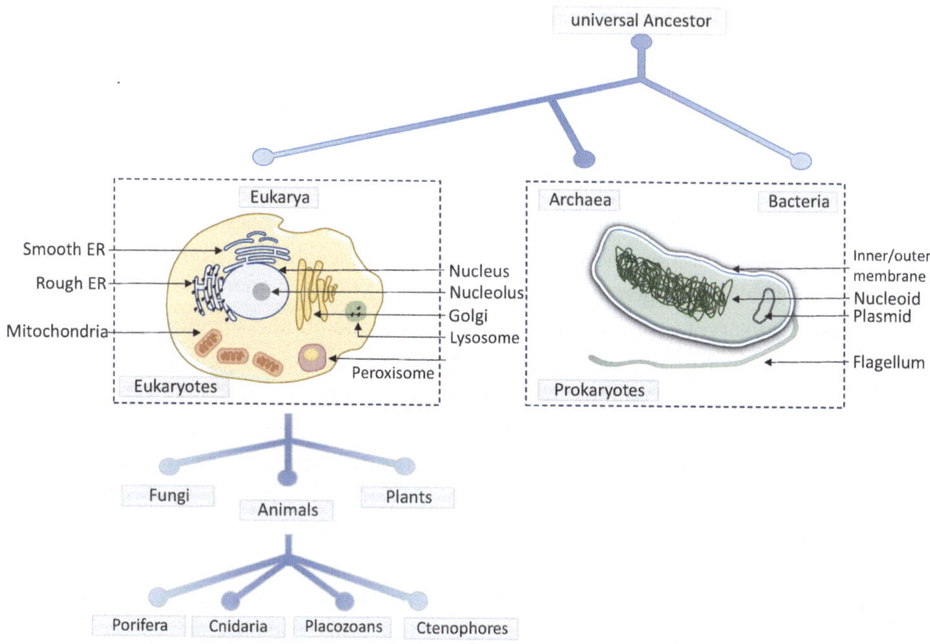

Fig. 1.5 Evolutionary tree of life considering universal ancestor cell with schematic representations of Eukaryotes and Prokaryotes. Cell organelles and cellular structures are indicated by thin black arrows. The tree of life is indicated by a division into Fungi, Animals and Plants. On the animal tree only the four phyla at the basis (porifera, cnidaria, placozoans and ctenophores) are shown

different shapes including coccus and rod-like. *Archaea* are thus similar to bacteria; however, there are significant differences. *Archaea* move very fast using several flagella and the chemical composition of their cell membrane is unique. Instead of a lipid bilayer, it is composed of a monolayer (see Chap. 3 for explanation). Moreover, the composition of their cell walls also differs from that of bacteria. It is composed of different polysaccharides, proteoglycans, and, sometimes, only of proteins; in addition, pseudomurein can be present and some *Archaea* lack a cell wall altogether. Therefore, *Archaea* earn their own domain in the three-domain classification (Fig. 1.5). They include so-called "extremophiles" which live in very hostile environments, e.g., in hot springs at temperatures up to 110 °C, in glaciers, or at high salt concentrations in the Dead Sea. *Archaea* produce methane gas. Residing in the gut of mammals, they reduce CO_2 to methane, a process supporting the digestion of plant cellulose by ruminants. Especially for this property, *Archaea* are increasingly explored for biotechnology. Because they tolerate nonphysiological temperatures very well, they can be cultured under conditions where contamination with other microorganisms can be excluded (Pfeifer et al., 2021).

Bacteria have a cell wall and a plasma membrane. As in *Archaea*, their DNA is surrounded by proteins but is not enclosed in a membrane; it forms a so-called nucleoid—a nucleus-like structure. Bacteria cell walls are composed of murines (see Chap. 3) and show extrusions called pilei and flagella for moving around (Fig. 1.5). Bacteria are best known as pathogens, but this is only true for a very small number including *Pneumococcus* responsible for pneumonia, rod-like bacteria such as *Anthrax* causing lung disease and being quite infamously associated with chemical warfare experiments, and the spirochaete *Treponema* causing syphilis. Many bacteria, however, live in our body essentially as symbionts. They provide a microbiome which is indispensable for our health, may it be in the digestive tract or our mucous epithelium (Gomaa, 2020). Furthermore, we use bacteria in many traditional ways to produce food, such as yoghurt and cheese. Plants also live in symbiosis with bacteria. Legume family plants, for instance, benefit from rhizobia, bacteria that are capable of nitrogen fixation (Lindstrom & Mousavi, 2020).

In contrast, eukaryotes do have nuclei and further membrane-surrounded cell organelles like mitochondria, lysosomes, endoplasmic reticulum, Golgi, and peroxisomes. Eukaryotes are represented on three branches of the tree of life: fungi, plants, and animals (Fig. 1.5). Fungi developed a large metabolic diversity and were able to metabolize almost everything, and plants developed chloroplasts and thus performed photosynthesis using carbon dioxide and the energy of light to produce high-energy organic material and oxygen, which is taken up by animals as food. Animals cannot use the energy of light. They essentially rely on taking up readymade high-energy organic products from other organisms. In this way, every branch of the tree of life is deeply connected with every other. Life depends on atmospheric CO_2, which is the essential building brick for all organic material produced by photosynthetic life forms. During glacial periods, CO_2 concentration went down to 180 ppm (parts per million, 0.0180%). Studies show that such low CO_2 pressure can impair photosynthesis rate, reproduction, and growth of plants and thus, also the functioning of ecosystems and agriculture (Gerhart & Ward, 2010). In green houses for large-scale food production, CO_2 is added up to 1000–1200 ppm.

In this textbook, we concentrate on animals. At the basis of the animal tree of life are four phyla of so-called simple animal forms (Fig. 1.5), including porifera (sponges), placozoans, ctenophores such as the comb jellyfish, and cnidarians represented by corals, jellyfish medusa, and hydrozoans; the latter often appear as colonial life forms (Fig. 1.5). It is not clear yet in which order they may have evolved; in any case, they are all sisters to a protozoan clade of choanocytes (Balavoine & Adoutte, 1998). All higher animals have a bilaterian symmetry with three body axes including an anterior–posterior, a dorsal–ventral, and a left–right axis (Fig. 1.6). They are divided into protostomia and deuterostomia according to the position of the mouth in the early embryo. Protostomes include nematodes and insects, and deuterostomes include sea urchins, and all vertebrates.

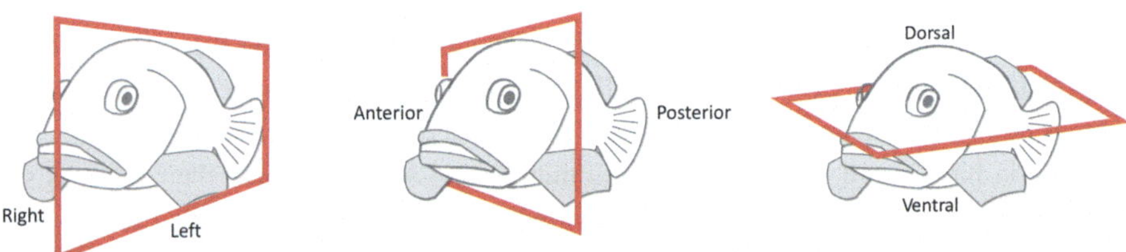

Fig. 1.6 Graphical representation for axis specifications in animals including anterior–posterior, dorsal–ventral, and left–right axes

Multicellular animals (metazoans), in contrast to single-celled life forms (such as protozoans), developed a number of novelties on the cellular level. These are epithelial cell connections, a three-component cytoskeleton, an extracellular matrix with its major component collagen, connections from the cell into the extracellular matrix, an elaborate molecular system for intercellular communication, and a nervous system.

Cells in multicellular organisms are specialized for specific functions and they come in different shapes and sizes. For instance, liver cells synthesize proteins for blood clotting, enzymes needed for storing and metabolizing nutrients and degradation of toxins, and many others. Muscle cells are packed with muscle fibers consisting of actin and myosin. Nerve cells produce proteins needed for neurotransmitter synthesis, synaptic structures, ion channels, and many more. Of interest for dentists are odontoblasts (Fig. 1.7).

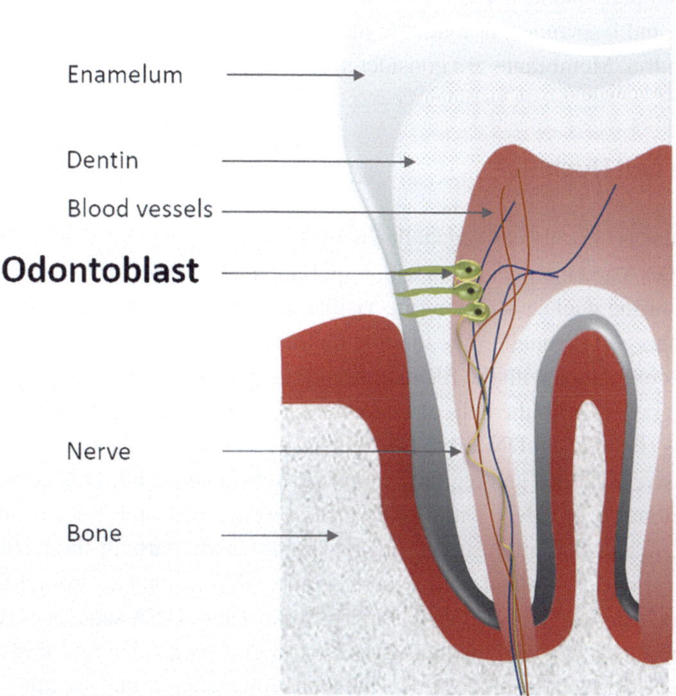

Fig. 1.7 Schematic representation of adult human tooth. Enamelum, dentin, blood vessels, nerves, and bone are indicated by arrows, and odontoblasts are highlighted

Two cell types are associated with human teeth. These are odontoblasts and ameloblasts (Fig. 1.7). Odontoblasts are palisade-shaped cells lining the dental pulp and secreting the extracellular matrix comprising our teeth. They live for the whole life of any of our teeth and allow formation of dental layers as well as repair. Moreover, they are connected with an excessive network of nerve fibers and therefore also act as a sensory organ to transmit temperature, mechanical, or chemical signals (Fig. 1.7). Ameloblasts form a densely packed sheet of cylindrical cells covering the spaces on top of the dentin layers in developing teeth. They secrete enamel and form the enamel ribbons. Once the tooth breaks through the pulp, these cells are sloughed off. We will refer to some of the very specific functions of tooth cells in the appropriate chapters.

1.2 Journey into the Cell: Cell Organelles

Cell organelles are membrane-surrounded cellular compartments in eukaryotic cells and include the nucleus, the endoplasmic reticulum (ER), the Golgi apparatus, mitochondria, lysosomes, and peroxisomes. The membrane can be a single membrane, as in ER, Golgi, peroxisomes, and lysosomes, or a double membrane as the ones surrounding the nucleus and mitochondria. Membranes are considered in detail in Chap. 3.

1.2.1 The Cell Nucleus

The cell nucleus is the only cell organelle visible by light microscopy (Fig. 1.2). It contains the genomic DNA and many proteins with important nuclear functions. In electron microscopy, details of the nuclear structure are visible including the nuclear double membrane surrounding this cell organelle. The inner membrane encloses the nucleus while the outer membrane is continuous with the ER membrane system (Fig. 1.8a, b). The genomic DNA is densely packed up with the help of DNA-binding proteins. Fluorescently labeled DNA in the nucleus (fluorescent DNA-dye DAPI is used) reveals specific patterns with regions that are intensively stained and regions, which are not stained at all. This demonstrates the specific distribution of DNA within the nucleus (Figs. 1.4 and 1.8c, d, animal cell). Furthermore, in electron microscopic images, high-density parts of the nucleus indicate the nucleolus (Fig. 1.8a animal cell), where certain DNA sequences from different chromosomes are clustering. It is also visible in Fig. 1.2a. These DNA sequences represent the regions of the DNA where ribosomal RNAs (rRNA) are encoded and transcribed. The nucleolus had also been pointed out in the light microscopic images after Richardson staining shown in Fig. 1.2 for plant and animal cells. Ribosomal subunits consisting of proteins and rRNAs are then assembled while still in the nucleolus and transported into the cytoplasm afterwards. Proteins involved in these processes can be detected in the nucleolus

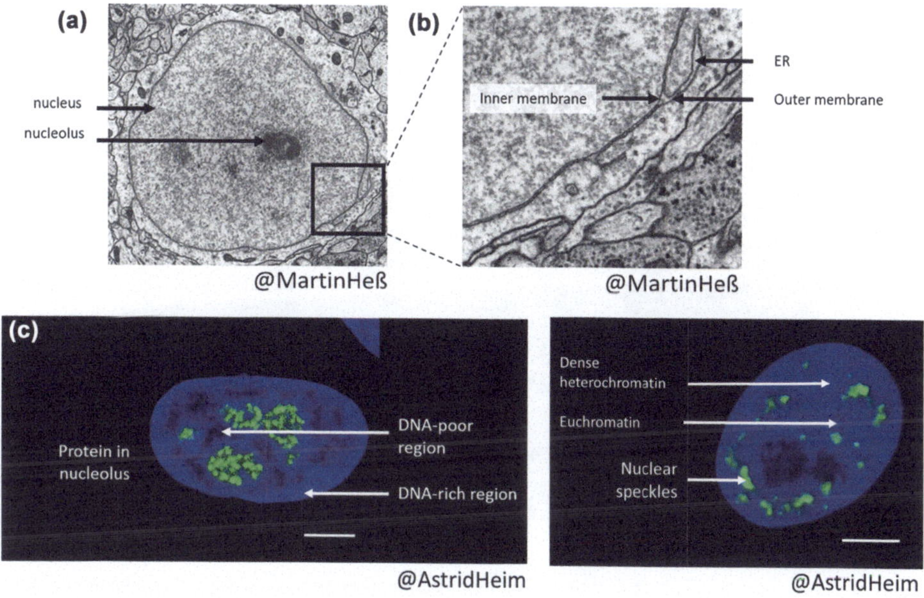

Fig. 1.8 (**a**, **b**) Electron microscopy of animal cells; (**a**) Detailed electron microscopy image of nucleus with nucleoli; (**b**) Enlargement of nuclear membrane showing the inner membrane that encloses the nucleus and the outer membrane of the nucleus, which is continuous with ER membranes; (**c**, **d**) Fluorescence microscopy of animal cell nucleus; (**c**) Nucleus is stained with DNA-specific dye (DAPI, blue), laser confocal microscopic image, dark regions are DNA-poor; light regions are DNA-rich, indicated by arrows. The nucleolar protein UBF is labeled with antibody (green), indicating the position of the nucleolus; (**d**) fluorescent laser confocal microscopic image of euchromatin and heterochromatin labeled with DNA dye. DAPI is shown. The nuclear protein SC35 was fluorescently labeled and appears in "speckled" pattern within euchromatin regions, C, D scale bars 5 nm

by immunofluorescence staining or by labeling with the green fluorescent protein, see later (for example, a protein called UBF as shown in Fig. 1.8c). Other patterns in the nucleus resemble speckles, which are not stained with the DNA dye, indicating that they do not contain much DNA, if any. Instead, they contain proteins like transcription factors, DNA polymerases, splice factors (see Chap. 2), and others (e.g., a protein called SC35 as shown in Fig. 1.8d). Nuclear speckles and the nucleolus therefore represent subnuclear compartments with specific functions.

Importantly and not visible at the resolution achieved in the electron micrographs in Figs. 1.8 and 1.9, both nuclear membranes are perforated by nuclear pores to allow molecular transport in and out of the nucleus. They are big enough to facilitate the transport of ribosomal subunits from their site of assembly in the nucleolus to the cytoplasm. However, these pores are not completely open and compound exchange requires a regulated transport system, which will be discussed in Chap. 5.

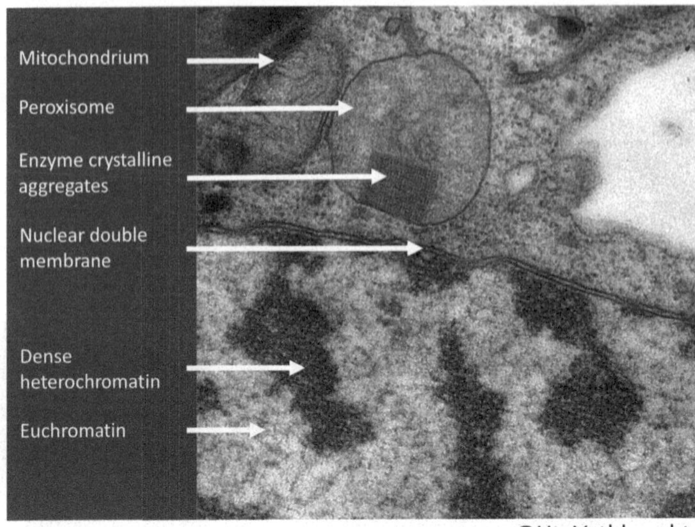

Fig. 1.9 TEM image of plant cell with nuclear double membrane, dense heterochromatin and euchromatin indicated by arrows, also mitochondrion, peroxisome with dense crystalline protein condensates

1.2.2 The Endoplasmic Reticulum (ER)

The ER is a network of membrane tubules passing through the whole 3D-room of the cell (Fig. 1.4a fluorescence microscopy, Figs. 1.3a, c and 1.8a, b electron microscopy). Moreover, we can distinguish between a "naked" part of this membrane system, called the smooth ER, and a part that is covered by ribosomes, the rough ER (Fig. 1.3c). Proteins are transported co-translationally into the ER, meaning that the ribosomes attach to the ER and deliver the newly formed proteins through a pore into the lumen of the ER, or into the ER membrane (Chap. 3). The smooth ER is free of ribosomes and serves lipid biogenesis and transport, Ca^{2+} storage, especially in the specialized ER of muscle cells (the sarcoplasmic reticulum, SR) to secure the high demand on Ca^{2+} ions for muscle contraction (Chap. 4).

1.2.3 The Golgi Apparatus

Relatively close to the nucleus in all cells, we find a membrane system consisting of several separate cisternae appearing as thin long tubes (Figs. 1.3b electron microscopy, and 1.4b fluorescent microscopy). This is the Golgi apparatus. It is not only a trading hub for vesicles but also a factory for post-translational modifications of proteins. Proteins that have entered the ER, with a few exceptions, are usually trapped in the membrane compartments of the cell and can only be transported within vesicles. Such vesicles bud off from membranes and fuse with other membranes. The Golgi apparatus receives vesicles from the ER on its cis side (the side facing the ER and nucleus). These vesicles fuse with the Golgi membrane and deliver their content into the Golgi cisternae, where it is usually modified, e.g., by proteolytic processing, glycosylation, and others. After passing the Golgi, proteins are again packed into vesicles, which bud off from the Golgi on the trans side (the side facing the cell membrane) (Fig. 1.3b). They are then transported to further membrane compartments, e.g., to lysosomes or into secretory vesicular compartments of a trans-Golgi network (TGN). From there, vesicles travel to the cell membrane and fuse with it to deliver transmembrane proteins into the cell membrane or they release their content (secreted proteins) into the extracellular space. Vesicles containing ER-resident proteins move by so-called "retrograde transport" back to the ER after being processed by the Golgi (see Chap. 3).

1.2.4 Mitochondria

Mitochondria are often called the "powerhouse" of the cell. Here, ATP is produced from organic compounds by the enzymes of the respiratory chain (will be explained in more detail in Sect. 1.3.2 and Fig. 1.21b). The enzymes of the citrate cycle and for beta-oxidation of fatty acids are also resident in mitochondria. Mitochondria have a size of 0.5–1 µm, and thus, they are smaller than the resolution limit of light microscopy; however, they can be visualized by fluorescence microscopy (Fig. 1.4c). By EM, their outer membrane and a highly extended inner membrane with large infoldings called cristae become visible (Fig. 1.3c). Due to the thin EM sections and the harsh fixation procedures applied to EM, it was thought for decades that mitochondria were single egg-shaped entities. However, modern investigations, especially live images of cells after fluorescence labeling in fluorescence microscopy, have shown that mitochondria form tubules extending throughout the whole cell (Figs. 1.4c and). These tubules are always engaged in forming and breaking up, which we call fusion and fission, respectively. In this way, they are highly adaptable to the physiological state and need of the individual cells. Figure 1.10 shows images of fluorescently labeled mitochondria in a dividing yeast cell with two daughter cells already visible and in cells of a simple cnidarian organism (*Hydra vulgaris*), which were manipulated genetically by introducing or not introducing a specific protein (mitofusin) to produce either long tubules or fragmented mitochondria (Fig. 1.10), (Müller-Taubenberger et al., 2006). Mitochondria contain their own DNA and ribosomes and contribute to the production of proteins for certain subunits of the respiratory chain. Mitochondrial DNA can be visualized by using heavy metals and is found in a relatively small part of the mitochondrial matrix, where it is organized in nucleoids, similar to bacterial DNA.

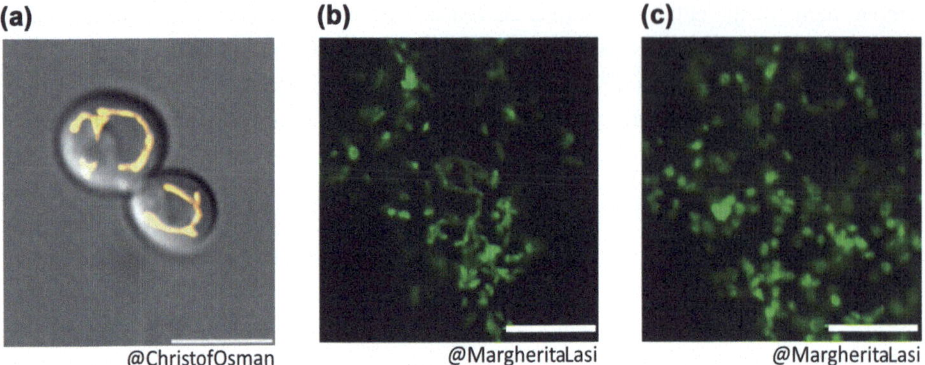

Fig. 1.10 (**a**) Light microscopic image of dividing yeast cell overlaid with fluorescence image of tubular mitochondria; (**b**, **c**) Laser confocal microscopic image of fluorescently labeled mitochondria in epithelial cell of *Hydra vulgaris;* Tubular in **b**; Fragmented in **c**; scale bar 5 nm

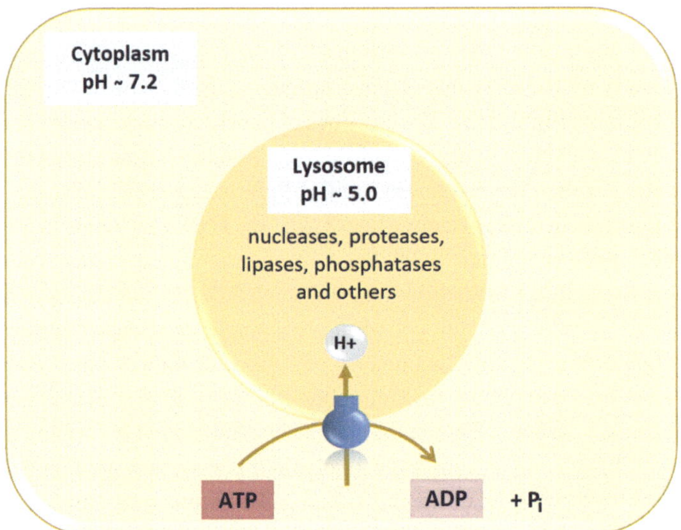

Fig. 1.11 Schematic representation of lysosome, pH-values for cytoplasm and lysosome and proton pump transferring protons into lysosome by using energy of ATP are indicated

1.2.5 Lysosomes

In lysosomes, all macromolecules of the cell and even small cell organelles can be degraded. Lysosomes are filled with lytic enzymes including nucleases, proteases, lipases, phosphatases, and others. Lysosomal enzymes have a pH optimum of 5.0, in contrast to the pH in the cytoplasm of 7.2. This is achieved by the activity of a proton pump on the lysosomal membrane, which enriches the lysosomes with protons (Fig. 1.11). The acidic pH optimum for lysosomal hydrolases renders them inactive at pH 7.2, should they accidentally reach the cytoplasm, e.g., when lysosomes are damaged. Cell organelles can also be degraded in the lysosome. In this case, a secondary lysosome is formed when the primary lysosome engulfs an organelle. Lysosomes can be made visible with fluorescent dyes that are enriched at low pH, for instance, acridine orange or lysotracker.

1.2.6 Peroxisomes

Peroxisomes are important organelles in the cell for detoxification. They contain enzymes for catalyzing oxidative degradation of fatty acids, ethanol, and other harmful compounds. They convert hydrogen peroxide into water using the enzymes peroxidase and catalase. Thus, they are also responsible for degradation of reactive oxygen species, which can otherwise propose a danger to cells due to their high reactivity. Peroxisomes contain enzymes in excessive amounts. Enzymes can reach such a high concentration that they form crystalline aggregates. This is shown for a cell in the leaf of the plant *Deschampsia caespitosa* in Fig. 1.9.

1.2.7 The Cytosol

The liquid content of the cell outside of cell organelles is the cytosol. It consists of about 85% of water. However, colloidal-osmotic pressure gives it the character of a gel rather than water. This is due to the presence of large amounts of densely packed macromolecules. These macromolecules include mainly proteins (15%) or protein–RNA complexes including ribosomes and proteasomes, but also lipids, carbohydrates, and nucleic acids, especially mRNAs, which serve the ribosomes as matrix to synthesize proteins (McGuffee & Elcock, 2010). The cytosol has a neutral pH and is able to buffer changes in proton concentration that may occur during metabolic reactions occurring in the cytosol, including glycolysis (Chap. 2). Moreover, the cytosol is permeated by the cytoskeleton, which consists of 3-D-networks of actin filaments, microtubules, and intermediate filaments, all of which shape the cell and support important cellular functions including mitosis, vesicle transport, intercellular and matrix attachment, and cell motility.

1.3 How Can We Investigate Cells?

Our current understanding of molecular cell structure and function has come from three lines of investigation methods. These include cytology, biochemistry, and genetics. Cytology analyzes the structure of cells and cellular components using microscopes and histological staining techniques. As we have seen before, there are some methods to non-specifically label cellular components with histological dyes; however, in recent years, much progress was made in specifically labeling single proteins of a cell. Biochemistry separates cells and tissues into their single molecular components and analyzes those, using centrifugation, chromatography, electrophoresis, and mass spectrometry. Finally, when it comes to understanding cellular functions, genetics enters the game. Geneticists search for correlations between genotype and phenotype by studying naturally occurring or artificially created mutants. Important techniques for such studies include DNA sequencing, cloning, and genetic manipulation. Often, model organisms are used. As all life on earth has many molecular functions in common, these model organisms range from bacteria, such as *Escherichia coli*, via fungi, e.g., budding yeast to plants like *Arabidopsis* and animals including invertebrates, such as the nematode *Caenorhabditis*, and the insect *Drosophila* and vertebrates like zebrafish and mouse. For investigations on mammals, cells are cultured outside their organism in tissue culture plates (Fig. 1.12).

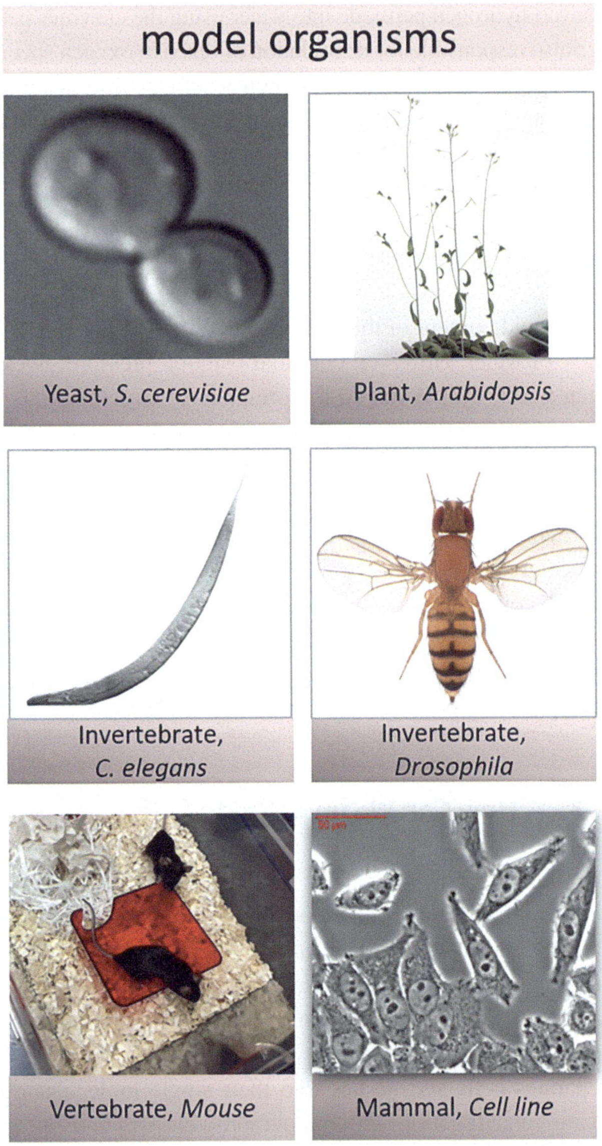

Fig. 1.12 Examples of model organisms used in cell biology research. They include the yeast *Saccharomyces cerevisiae (S.cerevisiae)*, the plant *Arabidopsis thaliana (Arabidopsis)*, the invertebrate *Caenorhabditis elegans (C.elegans)*, the invertebrate *Drosophila melanogaster (Drosophila)* in addition to *Mus musculus* (mouse) and a mammalian cell line; the images are labelled with the names that are commonly used in research laboratories (Photographs were provided by Christoph Osman, Tamara Mikeladze-Dvali, Cordelia Bolle, Heinrich Flaswinkel LMU Munich, and Anne Classen, Freiburg)

In the following, the historic development and the essence of methods of cytology, biochemistry, and genetics providing the basis of our current understanding of cell biology are described.

1.3.1 Cytology

Figure 1.13 indicates the limits of resolution for our own eyes and for different microscopic techniques. While light microscopy is limited by the wavelength of the light and cannot exceed a resolution of 200–400 nm, electron microscopy, by using electron beams, achieves resolutions down to 1 nm. In transmission electron microscopy (TEM), an electron beam emitted from a tungsten electrode penetrates the specimen and creates an image on the detector. Electromagnetic condensers and objective lenses direct the electrons on their way to the detector. The specimen is usually fixed and contrasted with chemicals (e.g., aldehydes and heavy metals like osmium or uranium) and embedded into an epoxy resin before being cut into ultrathin sections (such an image is shown in Fig. 1.3). In scanning electron microscopy (SEM), the electron beam is focused onto the specimen and secondary electrons backscattered from the gold-coated sample surface are detected. This shows reliefs of the surface of the observed object (Fig. 1.14). There are recent innovations in EM, e.g., block face scanning EM. Here, the specimen is also fixed, contrasted with heavy metals and embedded into a resin. Then, ultrathin slices are cut off and now the surface underneath the cut is scanned and the backscattered electrons are detected (such images are shown in Fig. 1.3). This is done slice after slice, with the first advantage that scanning is carried out on the specimen within the resin, which avoids distortion and loss of planes, and secondly, the scans after every slice can be combined into 3D images (Denk & Horstmann, 2004).

Light microscopy can be improved by using optical tricks, such as phase-contrast and Nomarski optic. In phase-contrast microscopy, rings are placed into the light path for structured light diffraction. This increases the contrast of unstained objects (e.g., see

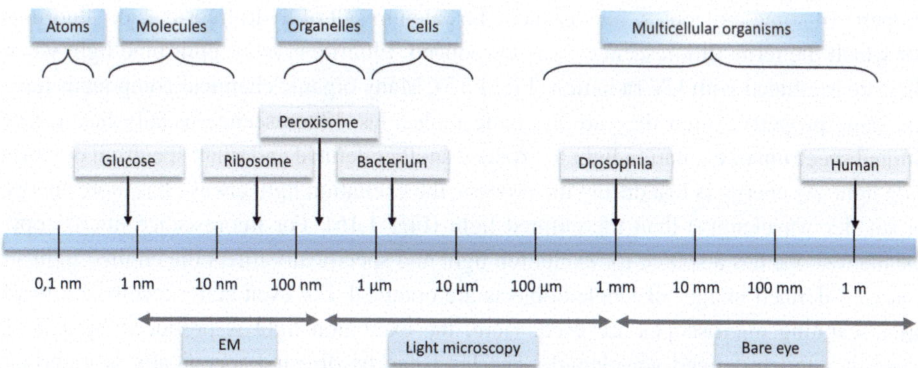

Fig. 1.13 Limits for optical resolution in light and electron microscopy in comparison with bare eye

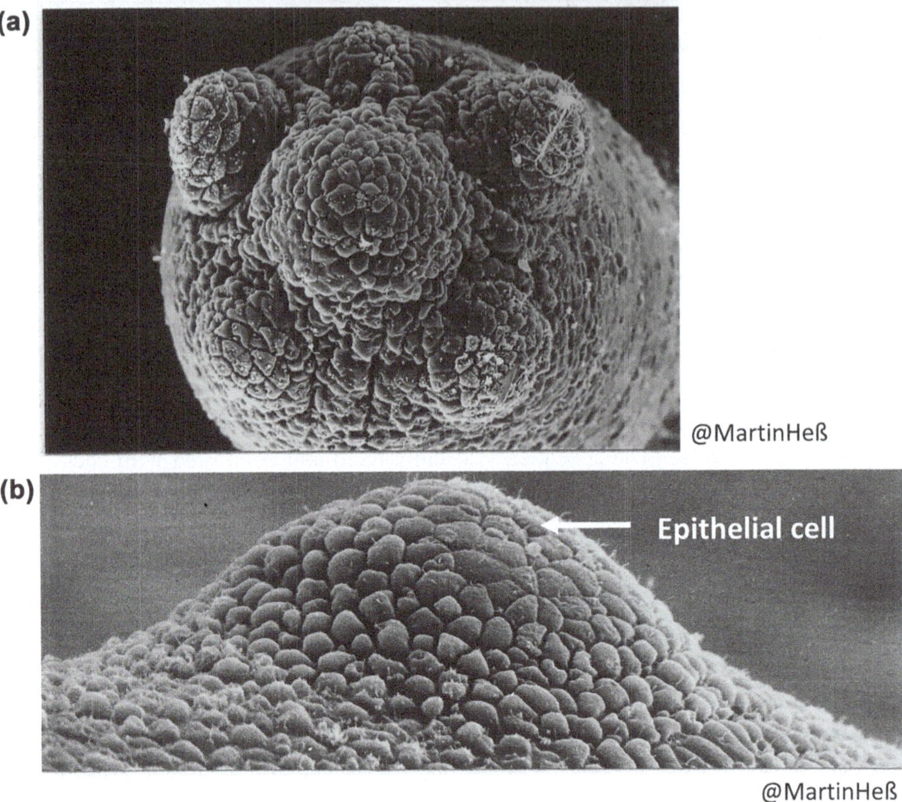

Fig. 1.14 Scanning electron microscopic images; (**a**) Young budding *Hydra* polyp with mouth cone and tentacle buds; (**b**) Very early *Hydra* bud with epithelia cell as indicated

Fig. 1.12, mammals, cell lines). Nomarski-Optic uses polarized light and delivers images of the surface relief of objects. Moreover, molecules can be visualized by fluorescent labels. In this case, cellular components are tagged with fluorophores. Fluorescence is a specific property of some molecules, for instance, the calcium salt of hydrofluoric acid, "fluorite," for which the term "fluorescence" was introduced. Fluorite crystals emit blue light when they are irradiated with UV radiation (Fig. 1.15). Many organic chemical compounds have the same property; often they are aromatic molecules. Fluorescence usually has a very limited spectrum of excitation light to produce another defined and small spectrum of emission light. As energy is lost during the process, the excitation light always has more energy (a smaller wavelength) than the emitted light (Fig. 1.16). For fluorescence microscopy, the microscope has a source for excitation light and specifically filters the emitted light so that well-defined images of labeled objects are obtained. For even better results, confocal laser scanning microscopes are used. Here, the excitation light is produced by a laser beam at a much-defined wavelength. The laser can be directed specifically at a certain focal plane of the object. In this way, spacious objects can be scanned plane by plane,

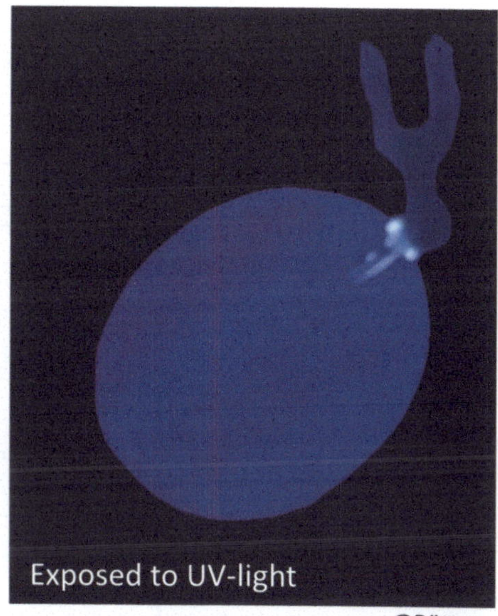

Fig. 1.15 Piece of jewelry, fluorite, imaged with visible light (left-hand side) and under UV (right-hand side)

thus significantly reducing light scattering and revealing very sharp images (e.g., see Fig. 1.4). In recent years, fluorescence microscopy has benefited from the development of so-called super-resolution microscopy. This involves several techniques to decrease the resolution below the diffraction limit including "Stimulated emission-depletion (STED)" fluorescence microscopy, where the excitation laser beam is followed by another, which dampens excitation in its surrounding, thus making objects visible, which are otherwise obscured by the emitted light wave. This technique was rewarded with a Nobel Prize in Chemistry for super-resolved fluorescence microscopy to Eric Betzig, W.E. Moerner, and Stefan Hell in 2014. In another method, the object is illuminated by a stripe pattern with the smallest distance close to the diffraction limit. The emitted light *interferes* with the properties of the object (Moire effect). It is then processed together with images obtained from different angles (interference-based super resolution by interference microscopy, 2D- or 3D-SIM); for more detailed explanations see, Schermelleh et al. (2019). We show an example of 3D-SIM imaging in comparison with images of similar objects obtained by laser confocal scanning microscopy (Fig. 1.17).

Only few molecules in the cell have their own fluorescence and can be observed by fluorescence microscopy, without artificial fluorescence labeling. One example is chlorophyll. It absorbs visible light of wavelengths between 400 and 700 nm and emits part of it as heat or red fluorescence. The most common methods for fluorescence labeling of specific cellular molecules are based on the use of antibodies (Fig. 1.18a). Antibodies are produced by laboratory animals after they have been injected with cellular molecules

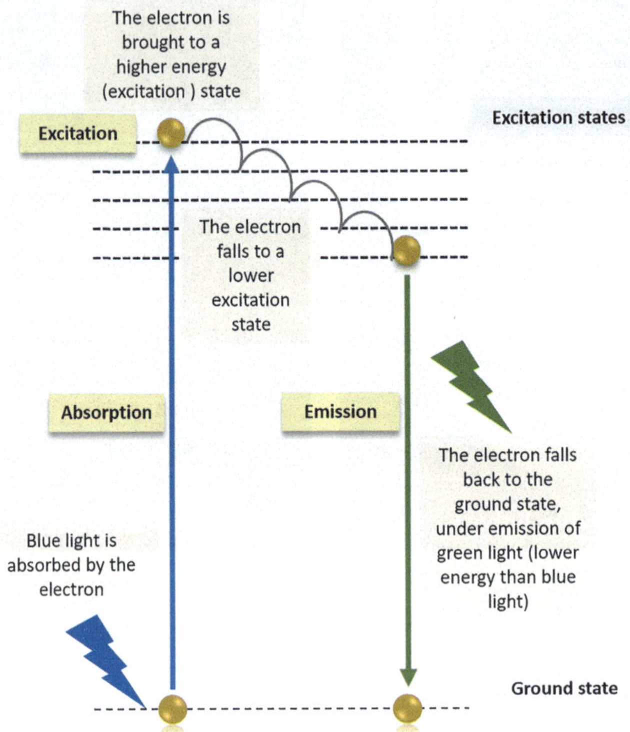

Fig. 1.16 Fluorescence is a property of certain materials, which emit light after excitation. Excitation occurs with higher energy (shorter wave length) light than emission. After excitation, electrons are transferred to a high energy level, from where they decay quickly to an intermediate energy state. By emitting light (fluorescence), they slowly return to the ground state. The emitted light has lower energy (longer wave length) than the excitation light

(antigens) as part of their adaptive immune response toward "intruders." Antibodies are produced by B cells of the blood system. Such B cells that are activated by binding of an "intruding protein" grow out and produce antigen-specific B-cell clones. Antibodies of different clones can be directed against specific regions or "epitopes" of the antigen. As a rule of thumb, it is expected that a protein has more possible epitopes, the bigger its molecular mass is. Thus, for a bigger protein, several different B-cell clones might grow out, producing several different antibodies, each of which recognizes a different part (epitope) of the protein. Thus, the antibodies are "polyclonal". They are secreted into the blood and can be harvested from serum. Their amounts are limited because the animal will eventually stop producing them. Polyclonal sera are routinely generated in a variety of animals including rabbits, goats, donkeys, sheep, and horses.

The antibody-producing B cells cannot survive in culture over a longer period of time. In order to obtain single B-cell clones, which can be kept in stable long-term cultures and produce the desired antibody, the technique of "monoclonal antibody" production was developed

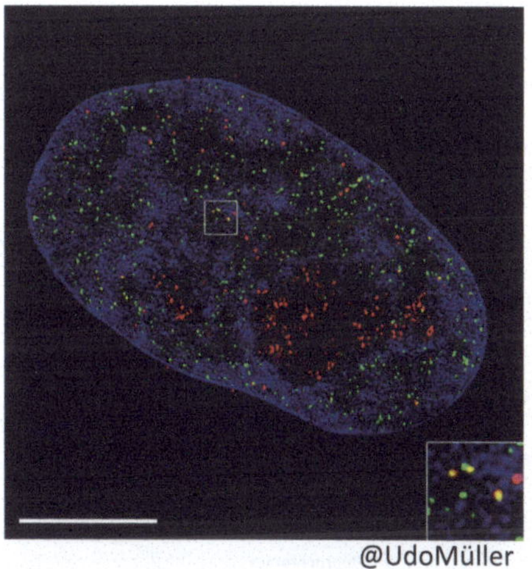

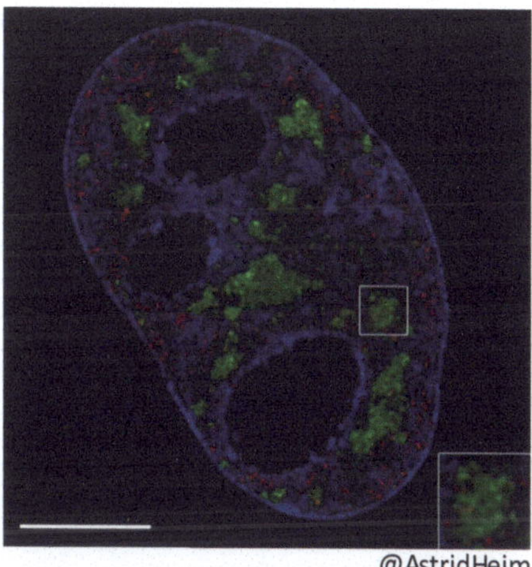

Fig. 1.17 Comparison of images of nuclear compartments obtained by 3D SIM microscopy (left-hand image) or laser confocal scanning microscopy (right-hand image). In both cases, optical sections are shown (meaning only one level of images), DNA shows blue fluorescence, nascent RNA (RNA that is just being transcribed, made visible by staining of incorporated 5′Fluorouridine with antibody) during transcription shows red fluorescence, and the protein U2AF65 shows green fluorescence. The inserts are magnifications of the white framed boxes. Scale bars 5 μm

and it was honored with a Nobel Prize for Milstein and Köhler in 1981. By this technique, antibodies producing B cells from immunized mice are isolated from the spleen and fused with immortal tumor cells derived from a myeloblastoma cell line. Successfully fused antibody-producing cells (called hybridoma cells) are then selected and tested for their specific antibody production. Positively tested hybridomas can be kept in culture indefinitely and secrete the desired, now monoclonal, antibodies into the culture medium. This technique is still used. Each monoclonal antibody only binds one epitope of the antigen.

Antibodies are immunoglobulins (Ig) mostly of the specific class IgG. These are tetramers, consisting of two "long" protein chains arranged together with two "short" amino acid chains. Two regions are distinguished within this protein complex, a constant region (Fc), which is essentially identical for all antibodies of a certain class, and a variable region (Fv), which develops a different amino acid composition in every evolving B-cell clone after antigen contact. Immunoglobulins can be directly labeled with fluorochromes, enzymes, or even gold particles for visualization. When they are applied to cells, the molecules they bind to become visible

- by light microscopy, when attached enzymes are treated with substrates yielding colored products by the enzyme reaction
- by "immunofluorescence," when the label is a fluorochrome
- by electron microscopy (EM) when the label is a gold particle

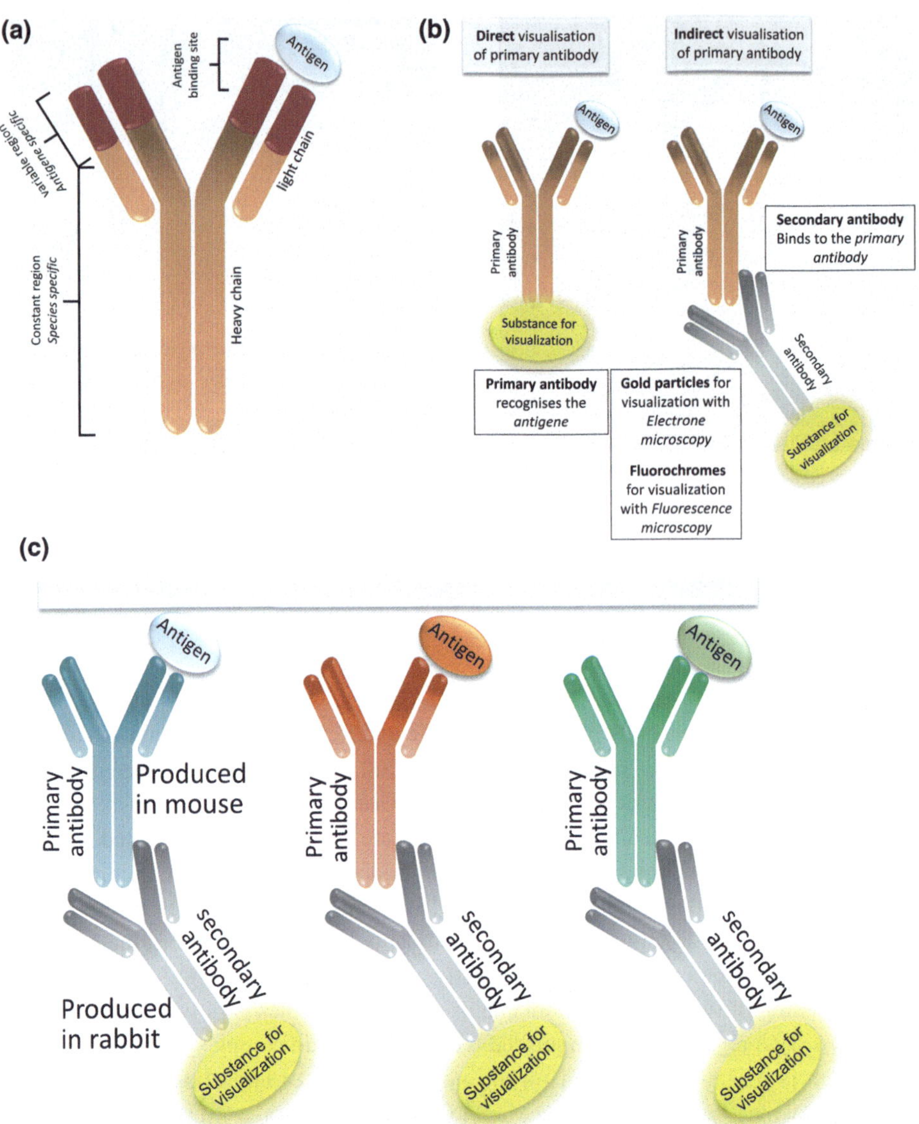

Fig. 1.18 (**a**) Schematic representation of IgG-class antibody. It is a tetramer consisting of four amino acid chains, two identical long chains, and two identical short chains. Variable regions are present on both, the light and the heavy chain, they are different for every B cell and can be further adapted to the antigen during B-cell maturation after antigen contact (V_H and V_L). Constant regions are also present on both chains, they are identical in primary structure for IgG-antibodies of one species, but differ between species (C_{H1}, C_{H2} and C_{H3} and C_L); (**b**) Schematic representation for direct labeling of antigen with labeled primary antibody and indirect labeling by using nonlabeled primary and labeled secondary antibody; labeling is indicated by yellow; antigen blue; (**c**) Labeled secondary antibodies bind to unlabeled primary antibodies. One secondary antibody binds to primary antibodies with different variable regions specific for different antigens

Commonly, indirect labeling methods are used. Here, the antigen-binding specific antibody (the "primary" antibody) is not labeled. Treatment of the sample with "primary" antibody is followed by a "secondary," now labeled antibody. This "secondary" antibody binds specifically to the constant region of the first antibody, so it binds to the antigen indirectly (Fig. 1.18b). Thus, a secondary antibody, produced in rabbits or horses after immunizing them with immunoglobulins from mice, can detect many different primary antibodies, which are produced by mice after injecting them with a specific antigen, e.g a protein, that we desire to investigate (Fig. 1.18b, c).

1.3.2 Biochemistry

In 1828, Friedrich Wöhler chemically synthesized urea, a product found in our urine, from inorganic ammonium cyanate. For this, he is often credited with providing the first evidence for a common chemistry of living matter and nonliving matter. When the cell theory was formulated, the prevailing view on life was that there must be a certain vital power "Lebenskraft," which is responsible for the creation, development, and death of all living things. Thus, life came into existence by "*generation aequivoca*" (Urzeugung) from nonliving and decaying organic material. It was Theodor Schwann who performed a number of experiments, which questioned this idea. He found that muscle meat would not decay when stored absolutely germ free. He used high temperatures to sterilize the meat and even the air and found that the meat would not change its appearance for 6 weeks. When he then allowed normal atmospheric air to enter his sample, it immediately started to rot. We know these procedures from methods for canning food. Schwann also looked at fermentation of wine and beer and by using his microscopes, he detected budding yeast in the beer mash and he described this as follows: "When the yeast was analyzed microscopically, the already-known grains were found, which make up the mash. However, I saw at first many of those connected in rows. In such a row, most of the time, some new rows branch sideways. Often, one can see between two grains a little bud sitting and founding a new row. And mostly on the last grain, a lengthy one was found. Summarizing, this whole thing has a great similarity with some fungi and is most likely a plant. Prof. Mayen, who was so kind to analyze this plant on my plea, came to the same opinion and said that it was only to debate whether it was an algae or a fibrillar fungus while favoring the latter due to a lack of green color." (*Bei der mikroskopischen Untersuchung der Bierhefe zeigten sich darin die bekannten Körnchen, welche das Ferment bilden; allein, ich sah zugleich die meisten derselben in Reihen zusammenhängen. Auf einer solchen Reihe stehen gewöhnlich ein oder mehrere andere Reihen schief auf. Häufig sieht man auch zwischen zwei Körnchen einer Reihe seitwärts ein kleines Körnchen aufsitzen, als Grundlage einer neuen Reihe, und meistens befindet sich an dem letzten Körnchen einer Reihe ebenfalls ein kleines, zuweilen etwas in die Länge gezogenes Körperchen. Kurz, das Ganze hat große Ähnlichkeit mit manchen gegliederten Pilzen und ist ohne Zweifel eine Pflanze. Herr Prof. Meyen, der diese Substanz auf meine Bitte ebenfalls zu untersuchen die Güte hatte, war ganz derselben Meinung und äußerte sich dahin, dass man nur zweifelhaft sein könne, ob es mehr für eine Alge oder für einen Fadenpilz zu halten sei, welches letztere ihm wegen des Mangels an grünem Pigment richtiger schien*) (Delbrück & Schrohe, 1904). The French engineer Cagniard-Latour (1777–1859) had made very similar observations by microscopic investigations of yeast and had submitted them to the Academy of Sciences in Paris in 1837 (Cagniard-Latour, 1838; Delbrück & Schrohe, 1904). He described the structure and behavior of yeast in a very detailed way and underlined his conviction that the fermentation process of vine was caused by a living organism: "All of those who are involved with vine fermentation, namely brewers and distillers know that their results despite the greatest care they take in carrying out their work end up to be extremely variable. Even those irregularities should strengthen the hypothesis that vine fermentation is carried out by living bodies, because 'who does not appreciate by how many ways such bodies can be influenced'" (Cagniard-Latour, 1838). The third person to microscopically analyze alcoholic ferments was the German primary school teacher Theodor Kützing. He also commented on the distinction of organic and inorganic matter with the following words: "I just want to remind you how we understand the term 'organic' in chemistry and in natural history. One has not come to a unified definition, because both sciences are following their own way. The physiologist, who has to consider all appearances of organic life, feels the need for a unification, because he cannot progress without it" (*"Ich erinnere hier nur daran, wie man den Begriff organisch in der Chemie und in der Naturgeschichte nimmt. Man hat sich hierüber noch nicht vereinigt, da beide Wissenschaften ihren Weg für sich gingen. Der Physiolog, der alle Erscheinungen des organischen Lebens zu berücksichtigen hat, fühlt das Bedürfnis einer Vereinigung um so mehr, weil er ohne sie nicht fortschreiten kann"*) (Kützing, 1837). The idea that yeast is a living organism and responsible for fermentation of grapes and crops was not received with great enthusiasm by other scientists in the mid-nineteenth century. In contrast, the three ingenious chemists Berzelius, Friedrich Wöhler, and Justus Liebig outrightly rejected it and continued to explain the fermentation process in purely inorganic (nonliving) terms. In his book, *Über die Erscheinungen der Gärung, Fäulnis und Verwesung*, Justus von Liebig explained how decomposition of living matter may create vibrations, which can induce processes like sugar fermentation. Eventually, Schwann and his colleagues were given justice by the work of Luis Pasteur, who very systematically analyzed fermentation processes and described different yeast species responsible for producing alcohol from sugar. He also performed metabolic experiments with yeast and is now considered the founder of fermentation chemistry (Delbrück & Schrohe, 1904). In 1907, Eduard and Hans Buchner showed that cell-free extracts from yeast could also do the job, which introduced the concept of biological catalysis, accomplished by ferments or enzymes, into biological sciences. "Zyme" in Greek means yeast and these discoveries founded the whole field of en*zym*ology. Hans Buchner died in 1902, and the Nobel Prize for their work was given to his brother Eduard in 1907. From 1905 onward, the single steps of sugar fermentation were discovered step by step by Arthur Haden and von Euler-Chelpin, which resulted in their receiving the 1929 Nobel Prize. Further steps were the description of lactose fermentation in muscle cells (or glycolysis), named after three researchers as the Emden–Meyerhof–Parnas pathway (also known as EMP pathway), the discovery of the citrate cycle named after its discoverer Hans Krebs as Krebs-Cycle, the discovery of the energy-rich universal compound ATP by Fritz Lipman and Karl Lohmann, and the resolution of the chemical steps of photosynthesis by Melvin Calvin, known as the Calvin cycle.

ATP plays a central role in energy metabolism. It is composed from an adenine base linked to the sugar ribose and phosphorylated on the C5 of the ribose carbon ring through a phosphor–ester bond (connecting the alcoholic OH of ribose to the phosphoric acid). The high-energy bonds are two phospho–anhydride bonds, which are formed between additional phosphates linked to the first one (Fig. 1.19). Explained in simple terms, the high energy contained in those phospho–anhydride bonds comes from the energy barrier to be overcome in order to join two negatively charged phosphate residues. Conversely, when these bonds are hydrolyzed, a lot of energy is released. More sophisticatedly, we would talk about a combination of charge repulsion and magnetic resonance stabilization. Therefore, ATP functions as currency for energy in the cell. Energy input by undergoing chemical reactions makes ATP, and ATP hydrolysis returns this energy (Fig. 1.19a, b).

Plants take the energy from light to form carbohydrates. Energy from carbohydrates is a central source for metabolic activity in plants and animals. Figure 1.20 puts the amount of energy stored in different molecular bonds in relation. Covalent bonds are such, where the two atoms involved share at least one electron pair, thus preventing their dissociation without addition of a reaction partner. Carbohydrate bonds represent very strong covalent bonds and store high amounts of energy. However, the phospho–anhydride bonds of ATP also fare quite well. Noncovalent bonds are formed by attraction of differentially charged ions, such as positively charged protons and negatively charged acidic groups or acidic and basic groups. They shift the charge between them, but can be easily separated and accordingly do not store a lot of energy. Van der Waals forces are active between hydrophobic compounds.

ATP is formed in our cells by both the process of glycolysis in the cytoplasm and the reactions of the citrate cycle (or Krebs cycle) in the mitochondria, where the end products of glycolysis are metabolized to produce electrons and CO_2. The electrons serve to produce a proton gradient across the inner mitochondrial membrane, which fires the mitochondrial ATP synthase. This enzyme produces the bulk of cellular ATP. Glycolysis is a

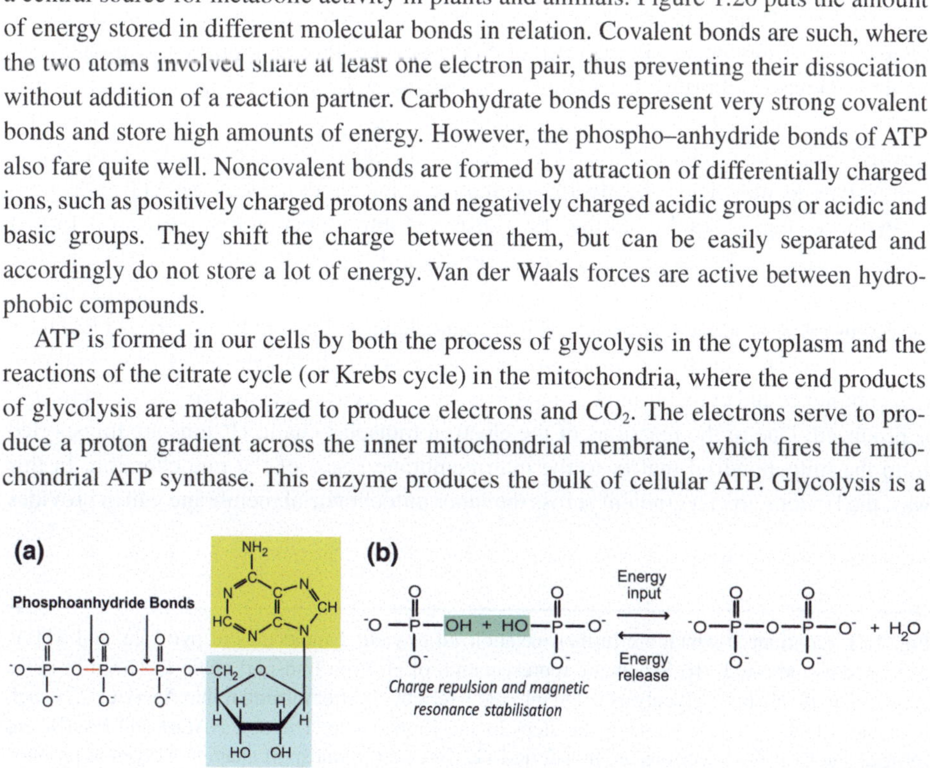

Fig. 1.19 (**a**) Structural formula of ATP, Adenin base (yellow box), Ribose (cyan box), and phospho–anhydride bonds (red) are indicated; (**b**) phospho–anhydride bond formation and hydrolysis

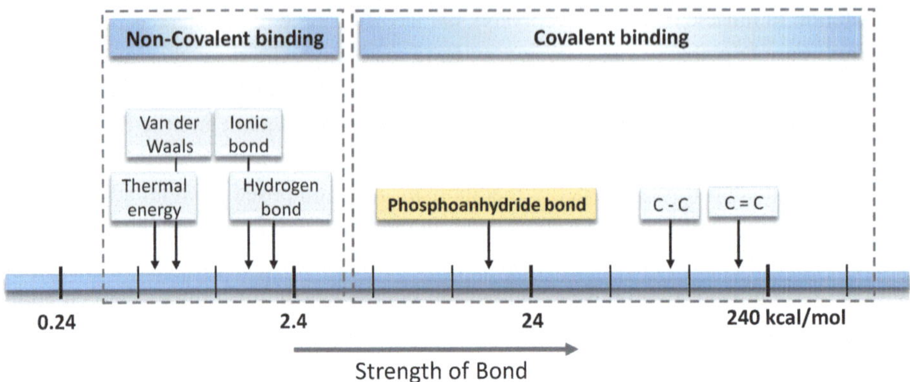

Fig. 1.20 Comparison of bonding energy for noncovalent and covalent bonds

highly complex process to break down the six-carbon chain of one glucose molecule into two molecules of pyruvate, which has a three-carbon chain. It involves ten different enzymes, ADP, ATP, NAD^+, and NADH and several steps (simplified in Fig. 1.21). Important steps in the glycolysis chain reaction include the cleavage of the glucose molecule (with 6 C atoms) and the phosphorylation of two aldehydes (with 3 C atoms each). These steps require energy in the form of ATP. For each glucose molecule, two ATP molecules are invested. Further steps include the oxidation, in which NAD+ is converted into NADH, and finally, the production of unphosphorylated pyruvate. The final steps of glycolysis require the input of four ADP molecules (for each glucose molecule) and produce four ATP molecules. Thus, the glycolysis chain reaction yields a net of two ATP molecules.

Pyruvate (Fig. 1.21a, b) can then be transferred into mitochondria, where it is further metabolized by the Krebs cycle to produce significantly more amounts of ATP (Fig. 1.21b). The cycle starts with the conversion of pyruvate into AcetylCoA and CO_2. The enzymes of the Krebs cycle produce CO_2, H_2O, ATP, H^+, and electrons (bound to NADH and $FADH_2$). The electrons are transferred along an electron transport chain, consisting of several protein complexes and cytochrome C, onto the terminal electron acceptor oxygen, where H_2O is produced. Due to the reactions of the electron transport chain, H^+-ions are transported from the mitochondrial matrix to the intermembrane space of the mitochondria. In this way, the H^+-ions erect a gradient across the inner mitochondrial membrane which provides

Fig. 1.21 (continued) which are further metabolized to yield 2 molecules of pyruvate, and ATP is generated as indicated; (**b**) Schematic representation of electron transport chain localized in mitochondria; pyruvate from glycolysis is fed into the Krebs cycle after turning it into AcetylCoA, which is oxidized to CO_2 by the 8 enzymatic steps of the Krebs cycle. Thereby, NADH and $FADH_2$ are formed, and their free electrons are transferred via the electron transport chain to oxygen to produce H_2O. Moreover, a proton gradient is established across the inner mitochondrial membrane, which fires the ATP synthase to produce ATP

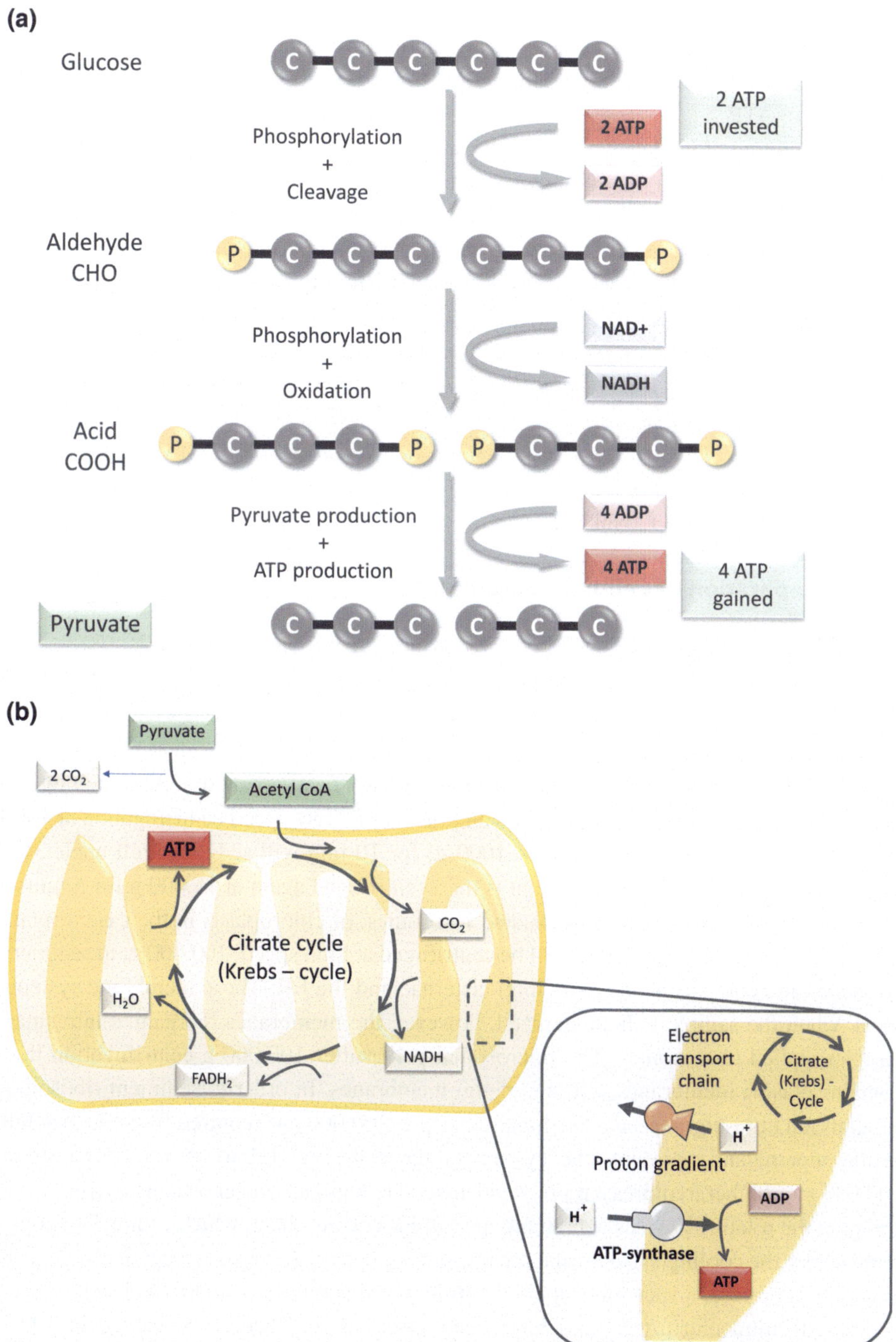

Fig. 1.21 Glycolysis and the Krebs cycle; (**a**) Highly simplified representation of glycolysis: in the first step, glucose (6-)carbon is cleaved and phosphorylated to yield two 3-carbon glyceraldehyde-phosphates, which are then stepwise phosphorylated and oxidized to yield 2 phosphoenolpyruvates,

energy for the ATP synthase to produce ATP (Fig. 1.21b). This process produces another 36 ATP molecules. Thus, the net energy production from one molecule of glucose after glycolysis had been two ATP; now, through the respiratory chain, this is increased to 38 molecules of ATP.

The beginning of nucleic acid research can also be dated back into the nineteenth century. Friedrich Miescher in Tübingen isolated a compound he called "Nuclein" from salmon sperm and from human pus, which he had received from a nearby hospital. He described his discovery in the following words: "After experiments on other tissues…it appears to me likely that a whole family of such, slightly distinct amongst each other phosphoric bodies will surface, which may deserve to be put equally side by side with proteins." (*"Nach Versuchen an anderweitigen Geweben…ist es mir nun wahrscheinlich, daß eine ganze Familie von solchen, untereinander etwas abweichenden phosphorhaltigen Körpern auftauchen wird, die vielleicht als Gruppe der Nukleinkörper den Einweißkörpern ebenbürtig gegenübergestellt zu werden verdient."*) (Miescher, 1871). Albrecht Kossel working in Berlin between 1879 and 1888 detected adenine and guanine in the nucleic acid part of "Nuklein."

1.3.2.1 Separation of Cellular Components

1.3.2.1.1 Centrifugation

Centrifugation allows to separate cellular components on the basis of their size, shape, and density. For separation of cell organelles by centrifugation, cells or tissues have to be lyzed and homogenized in such a way that the cell organelles are not destroyed. This requires the use of isotonic buffers to prevent bursting or shrinking of cell organelles. After homogenization, the lysate is cleared of any debris and then a first low-speed centrifugation at 1000 times the gravitational acceleration (= 1000 g) for 10 min will sediment cell nuclei, the largest cell organelles. The supernatant is then centrifuged again at 15,000 g for 5 min to sediment mitochondria, peroxisomes, and lysosomes, or chloroplasts in the case of plant cells. Eventually, the supernatant will be centrifuged at high speed, 100,000 g, to sediment all membrane compartments, like Golgi cisternae and the ER. These membrane systems break when the sample is homogenized; however, the membranes reorganize into small vesicles, called microsomes. The microsomal pellet after 100,000 g centrifugation thus contains plasma membranes, ER, and Golgi membranes. In order to sediment ribosomal subunits, an ultra-high-speed centrifugation step at 300,000 g is required. What is then left in the supernatant represents the cytosol of the cell (Fig. 1.22a). A centrifugation at 100,000 g or higher, requires a more sophisticated technique. The enormous friction from spinning the rotating devices containing the samples creates heat, which cannot be simply removed by the cooling systems that are present in all centrifuges to protect the biological material. Therefore, a vacuum is applied before the centrifuge is started. Such centrifuges are called "ultracentrifuges" and have been invented by Theodor Svedberg in 1920. In appreciation of his invention, we still refer to the "Svedberg" constant, a measure for the velocity required for a particle to sediment by centrifugation, multiplied with the

(a)

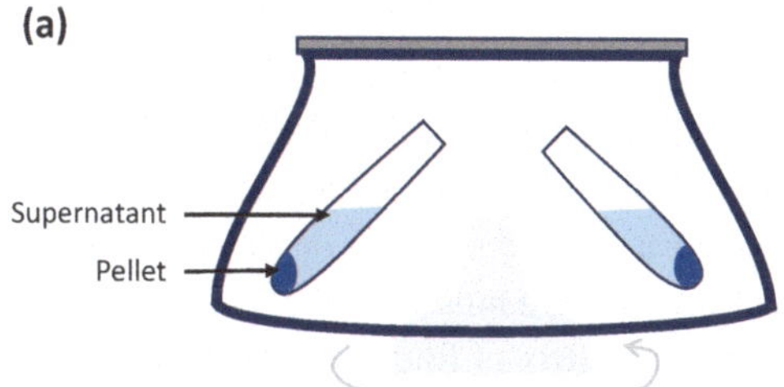

Centrifugation RCF (x g)	Content pellet
1.000	Cells and cell nuclei
15.000	Mitochondria, lysosomes, peroxisomes
100.000	„microsomes", polysomes, ribosomes
300.000	Ribosomal subunits

(b)

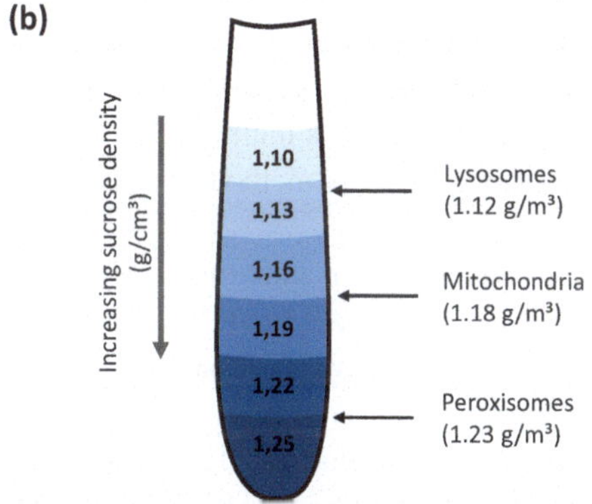

Fig. 1.22 Centrifugation; (**a**) Rotor with tubes, cellular components are sediment in pellets (dark blue) depending on acceleration force proportional to rotor speed given in revolutions per minute (rpm), leaving a supernatant fraction (light blue). Table indicates the content of pellets after centrifugation at different acceleration force values (RCF), which is given in multiples of the standard acceleration due to gravity on the Earth's surface (x g); (**b**) Centrifugation tube with preadjusted sucrose density gradient, and cellular components are enriched in regions of the gradient according to their density

centrifugation strength (a particle may sediment in a week outside the centrifuge at 1 g, but in 10 min, at 1000 g by centrifugation).

To further separate the cell organelles, one makes use of their differential density. With the help of a sucrose solution, a density gradient is established in a centrifuge tube. Density ranges from 1 to 1.25 g/cm^3. When a cell homogenate is layered on top of this gradient, by centrifugation, every organelle will be found enriched at the position of its own density in the sucrose tube. For example, lysosomes are found at 1.12 g/cm^3, mitochondria at 1.18 g/cm^3, and peroxisomes with the highest density at 1.23 g/cm^3 (Fig. 1.22b).

1.3.2.1.2 Chromatography

Further separation techniques used by biochemists include chromatography, electrophoresis, and mass spectrometry. For chromatography, the devices, matrices, and solvents may differ between the methods and techniques, but the principle is always the same: the different components of a substance are separated according to their properties in binding to a stationary phase (matrix) and dissolving in a mobile (solvent) phase. These separation methods are applied to identify and determine the concentration of the components of a mixture. For chromatography (chrome = color, graph = to write), substances are separated according to their differential solubility in a certain solvent, or in other words, according to how they distribute between a stationary and a mobile phase. For example, a mixture of differently colored powders is put onto a matrix of filter paper and then water is added as solvent. The capillary forces of the filter paper make the water travel away from its source and colors that are dissolved in the water will travel with it. The better they are dissolved, the further they travel and vice versa, the stronger their affinity to the paper, the less far they will travel. In this way, different components of the colored powder can be separated. In a chemistry lab, different 2 D matrices are used and a multitude of solvents are applied, including phenol, ethanol, or solvent mixtures. This method is called thin-layer chromatography (Fig. 1.23).

It is also possible to separate large amounts of compounds from a mixture in a preparative manner. Here, columns harboring large amounts of matrix are used. We distinguish between gel filtration, ion exchange, and affinity chromatography. The matrix is usually composed of agarose, a polysaccharide isolated from red algae. For gel filtration, substances are separated according to their size. Agarose is a porous material. Large molecules will not pass through the pores and therefore they have a short way from the top to the bottom of the column and they will be the first to arrive at the detector. Smaller molecules will pass through the pores and have a longer way, therefore they arrive at the detector later. Molecules of intermediate size will wander through some of the pores only, and

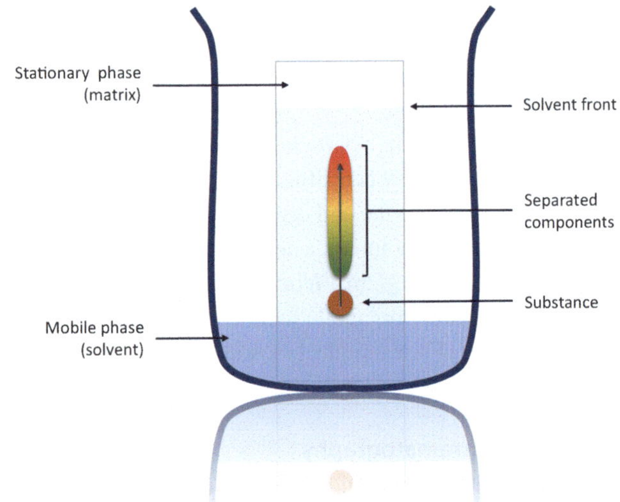

Fig. 1.23 Thin-layer chromatography; Mobile phase is indicated at the bottom of the vessel inside which the matrix (sheet of paper or other material) is placed. Due to capillary forces, the solvent travels upwards and takes substances with it, that are easily dissolved in the solvent, thus separating compounds of a mixture (from brown dot to colored mix)

accordingly, they will travel for times in such for large molecules and very small ones (Fig. 1.24). These columns are also suited to desalt a sample because the salt molecules will be the slowest, as they are delayed while travelling through every pore of the agarose.

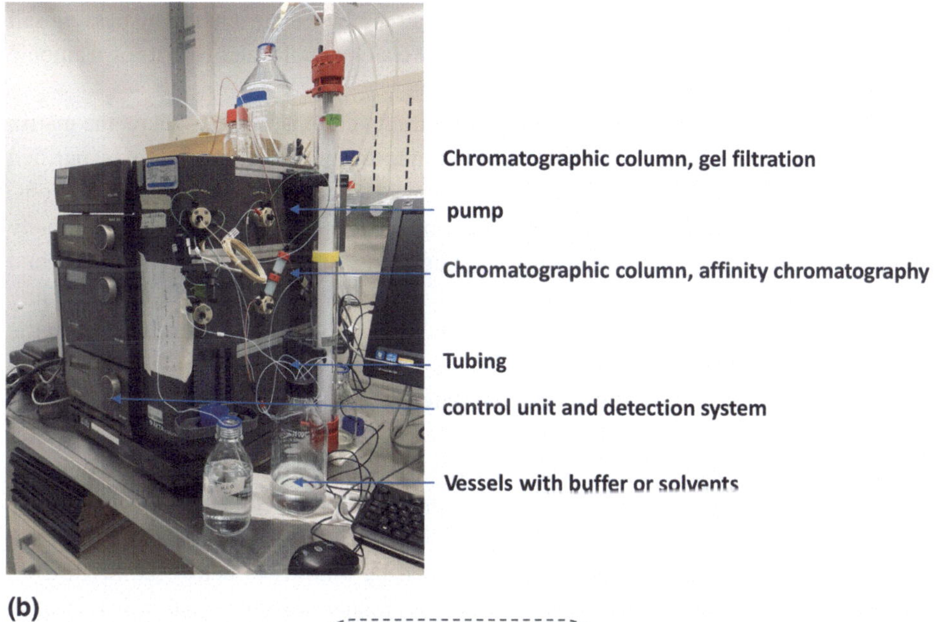

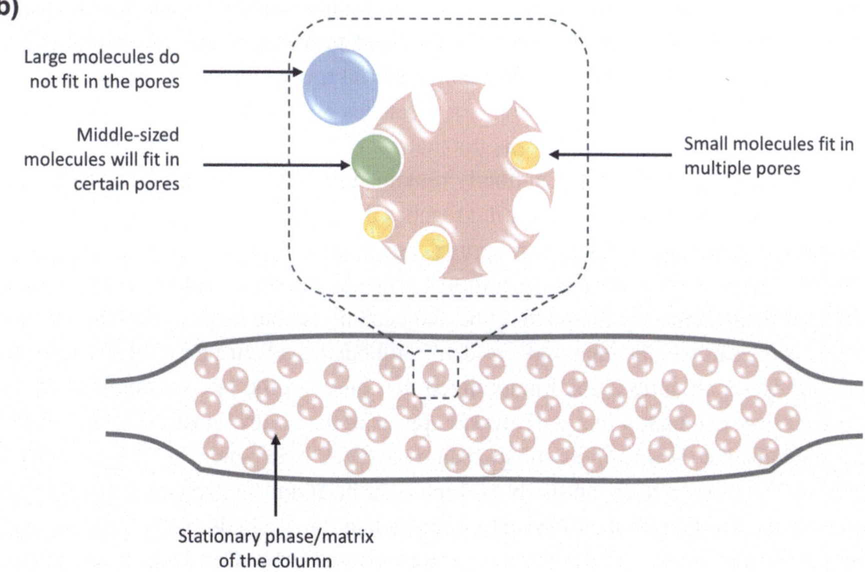

Fig. 1.24 (**a**) Standard FPLC system (*F*ast *p*rotein *l*iquid *c*hromatography); (**b**) Schematic representation of size exclusion chromatography, with enlargement of porous grains in stationary phase inside the column. Large molecules do not fit into the pores and therefore have a shorter way to move through the column arriving more quickly at its end than smaller molecules

In ion exchange chromatography, the matrix is covered with ions; these can be cations or anions. Molecules will bind to this charged matrix. Negatively charged molecules will bind to a cationic matrix and stay there, while positively charged molecules will be repulsed and run through the column. Vice versa, positively charged molecules will bind to an anionic matrix, while negatively charged molecules will run through. The bound molecules can then be released according to the strength of their binding by passing salt gradients along the column. The salt exchanges the charged molecules with its own anionic or cationic part.

A special case in column chromatography is affinity chromatography. Here, the matrix is covered with a substance, which specifically binds to something that should be purified from a mixture. This would be the case, for instance, for antibody–antigen interactions. Antibodies are normally very specific binders of their antigens. If we had an antibody binding the spike protein of coronaviruses, this antibody could be linked to the matrix and when spike protein passes the column, it would stick to this matrix, whereas everything else in the mixture was running through. The spike protein could then be eluted from the column with an acidic buffer. Another example is Nickel, which specifically binds to histidine residues. If proteins with histidine tags are passed through a column covered with nickel, they get stuck to it and later can be eluted with imidazole, which binds to nickel much stronger than histidine and replaces it, thus pushing the histidine tagged protein out.

A practical example for using thin-layer chromatography in combination with affinity chromatography is the so-called antigen quick test. Such tests are used, for instance, to indicate a pregnancy. Lately, such tests were used by billions to detect the SARSCoV 2-virus. Here, we have a plastic device with a precoated absorbent matrix inside it. A sample from the throat or nostrils is first solubilized in a buffer and then administered directly onto the adsorbent matrix. The soluble material travels through capillary forces to the other end of the device. While doing so, it catches a gold-labeled antivirus antibody that is solubilized when the sample runs the matrix. If the virus is present in the sample, the antibody binds to it; if not, the antibody remains unbound. In both cases, the labeled antibody travels in the mobile phase until it reaches the first line, where a second antibody is immobilized which also binds to the virus, but at a different position. If the sample contains the virus, a virus already in complex with the "golden" antibody will now be immobilized by the antibody on the first line and become visible by the gold label. If there is no virus in the sample, the unbound "golden" antibody travels further until it reaches the second line. Here, an antibody is immobilized that binds to the labeled antibody in any case. If the assay is intact, a line will always appear in the control position (Fig. 1.25). It appears obvious that the specificity of such an assay heavily depends on the specificity of the antivirus antibody. Often, antibody epitopes can be found in proteins other than the diagnostic ones (Böttger et al., 1995). Moreover, when the antibody binds to a virus protein that is present in all coronaviruses and not only in SARS-CoV2, such as the viral nucleocapsid (N) protein, the assay cannot point to the specific SARS-CoV 2 infection, but detects all coronavirus infections. Finally, especially in throat samples, due to food

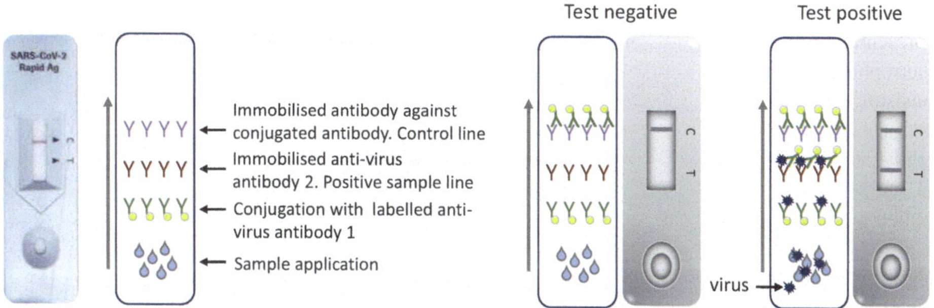

Fig. 1.25 SARS-CoV-antibody test as a combination of thin-layer chromatography and antibody affinity chromatography; antibody is indicated by Y-Symbol, Y with yellow dot shows labeled antibody, virus blue and sample light blue, as antigens nucleocapsid proteins were used; these are present in all coronaviruses and are not exclusive to SARS-CoV 2

intake (e.g., coffee or coke), other factors may interfere with the assay to make it appear positive, although no virus is present. However, the most important point is that there has to be strong correlation of such an essentially molecular assay with symptoms of the disease that it is supposed to detect. In the absence of symptoms the test may very well be meaningless.

1.3.2.1.3 Electrophoresis

The third method of separating biological material is electrophoresis where the compounds are separated in an electrical field. To separate DNA strands of different lengths from each other, agarose gels are used. They are usually put into a tray and have small slots cast in at one side, where the DNA is added. Then, a voltage is applied for a certain period of time between the cathode and the anode of a chamber in which the agarose gel is placed. DNA is negatively charged as we will see later. It thus will move within the gel in the direction of the positive pole, the anode (to avoid confusion: the anode is called anode, because anions, negatively charged ions, are attracted by it, it has a positive charge; conversely, positively charged ions or cations are attracted by the cathode, which has a negative charge). Within a certain time period, smaller DNA fragments will move faster toward the positive pole than larger ones (Fig. 1.26). This allows a very clear separation of pieces with a size difference under 100 base pairs. A DNA ladder with DNA fragments of known sizes is used to identify the size of the DNA fragments in the sample. The DNA is made visible in the gel by staining with suitable dyes, including ethidium bromide (Fig. 1.27).

Proteins can also be separated by electrophoresis. For this, usually acrylamide polymers are used. They form a net with different pore sizes dependent on the concentration of acrylamide and bis-acrylamide in the gel that is formed after polymerization of these compounds. Every protein has a net charge due to its composition of characteristic amino

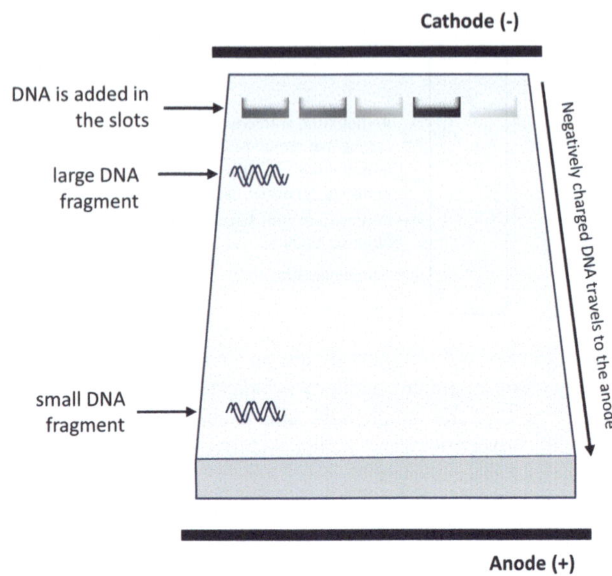

Fig. 1.26 Schematic representation of electrophoresis, sample consisting of DNA fragments is applied to small slots of an agarose gel, and voltage is applied with indicated polarity (+/−). Large fragments travel shorter distances than shorter fragments

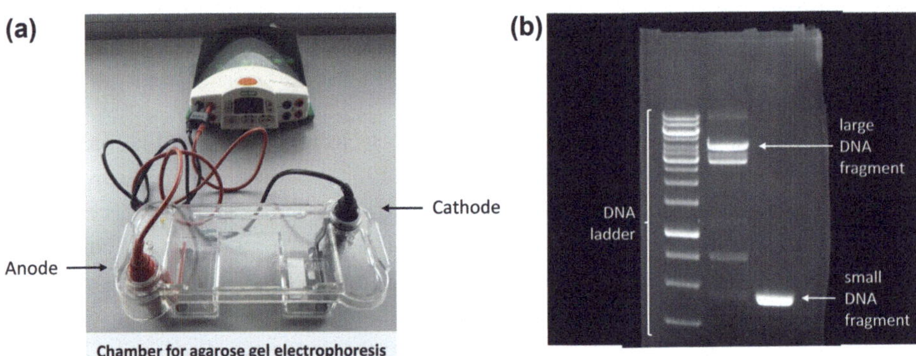

Fig. 1.27 (**a**) Standard power supply and gel chamber for agarose electrophoresis used for separating DNA fragments; (**b**) Typical result of DNA agarose electrophoresis; a gel is shown after running the electrophoresis and staining the DNA, first lane with DNA ladder containing 14 DNA fragments of known sizes, second lane sample containing several fragments, and third lane with one small fragment

acids that can be negatively charged, positively charged, or neutral. In native polyacrylamide electrophoresis, a protein will be attracted to the cathode when overall positively charged, and to the anode, when overall negatively charged. Similar to the DNA in the agarose gel, proteins will travel the faster in the gel, the smaller they are. In addition to the number of amino acids in the protein chain, the folding also has an impact on its spatial size. The more compact it is folded, the smaller it appears. In a native gel, the travel velocity of any given protein is therefore determined by its size and by its charge. In order to separate proteins solely by their size, SDS gel electrophoresis is used. Proteins are treated with a hydrophobic charged compound (sodium–dodecyl–sulfate), which binds to the surface of the proteins and denatures them while applying the negative charge of the sulfate group to them. This charge is so strong that it overrides the individual charge differences of the proteins in the mixture (Fig. 1.28a). Therefore, all proteins travel to the positively

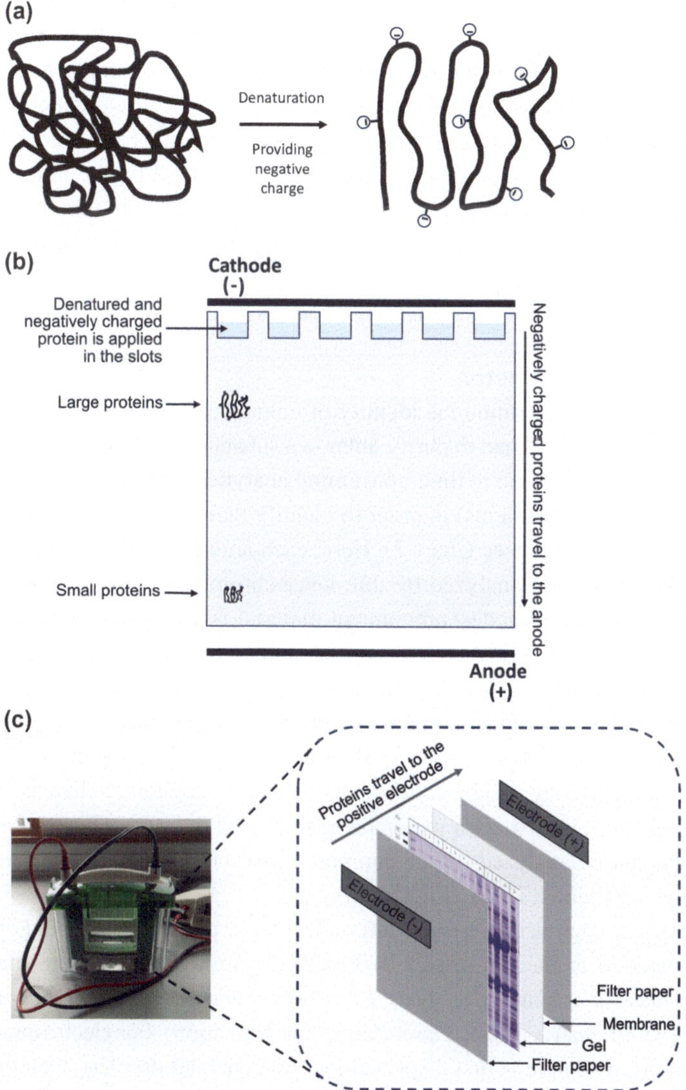

Fig. 1.28 (**a**) Schematic representation of denatured protein sample after treatment with SDS. Negative charges are indicated on the surface of the denatured protein; (**b**) Schematic representation of a polyacrylamide gel with cathode and anode, sample slots (light blue) and small and larger equally charged proteins; (**c**) Standard Western blot chamber connected to power supply (left), schematic representation of assembly of gel and nitrocellulose membrane between electrodes (right)

charged anode. Depending on the mesh size of the acrylamide gel, the proteins in the sample will now be separated solely according to their size determined by the number and molecular masses of their amino acid (see Fig. 1.28b). A so-called marker consisting of a mixture of proteins with known sizes can be applied along with the protein samples in order to estimate their molecular masses (Fig. 1.28b).

After separation, the proteins are usually analyzed further, for instance, by staining them with labeled antibodies in order to identify the proteins of interest. They cannot be treated with antibodies while on the poly-acrylamide gel but have to be transferred to a nitrocellulose membrane. This transfer is done by "Western blotting" (Fig. 1.28c). The membrane is put onto the gel and both are placed into an electrical chamber, voltage is applied, and the still negatively charged (SDS-bound) proteins travel toward the anode. When they arrive at the membrane, they are covalently bound by the nitrocellulose and cannot be removed anymore. If we perform this procedure with a mixture of mitochondrial proteins and we have an antibody, which specifically binds to the mitochondrial protein cytochrome C, we can tell the precise position of cytochrome C on the Western blot. The intensity of the band then tells us something about its amount.

1.3.2.1.4 Mass Spectrometry

A difficult matter is to determine the identity of unknown compounds in biological samples. Originally, researchers had to purify unknown substances to a high purity and in large amounts and then submit them to time-consuming analytical approaches (like Edman degradation and analysis for proteins) in order to identify their nature and in case of proteins, their amino acid sequences (see Chap. 2). Here, each amino acid had to be removed from the protein chain and then analyzed by thin-layer chromatography. A modern detection method, which needs only modest amounts of material is mass spectrometry. Mass spectrometry measures the masses of atoms and molecules and their number. Each atom has a defined mass as can be seen in the periodic system of elements (Fig. 1.29a). All compounds have to be brought to the gas phase before mass spectrometric analysis is possible. In the mass spectrometer, they are ionized and accelerated by voltage in order to "fly" through the machine. They will be deflected by an electromagnetic field and hit the detector at a position that relates to their mass/charge relationship (m/z). This is transferred into a signal which can be analyzed by the computer. Molecules with different masses show different "times of flight" (TOF), (Fig. 1.29b).

Large molecules such as biological macromolecules or parts of them could originally not easily be transferred to the gas phase. Two methods are now available to make proteins amenable for mass spectrometric analysis: Electrospray and matrix-assisted laser desorption ionization (MALDI, *m*atrix *a*ssisted *l*aser *d*esorption *i*onization). For electrospray ionization (ESI), the dissolved sample is first dispersed into fine charged droplets, then the solvent is evaporated, and ions are ejected from these highly charged droplets, and can then be separated in the mass spectrometer (Ho et al., 2003). For MALDI, a pulsed UV laser is directed onto the solid sample, which is embedded into a solid matrix. The laser energy resonates with the matrix leading to desorption of the sample from the matrix, whereafter it is ionized by laser excitation for analysis in the mass spectrometer (Hillenkamp & Karas, 1990).

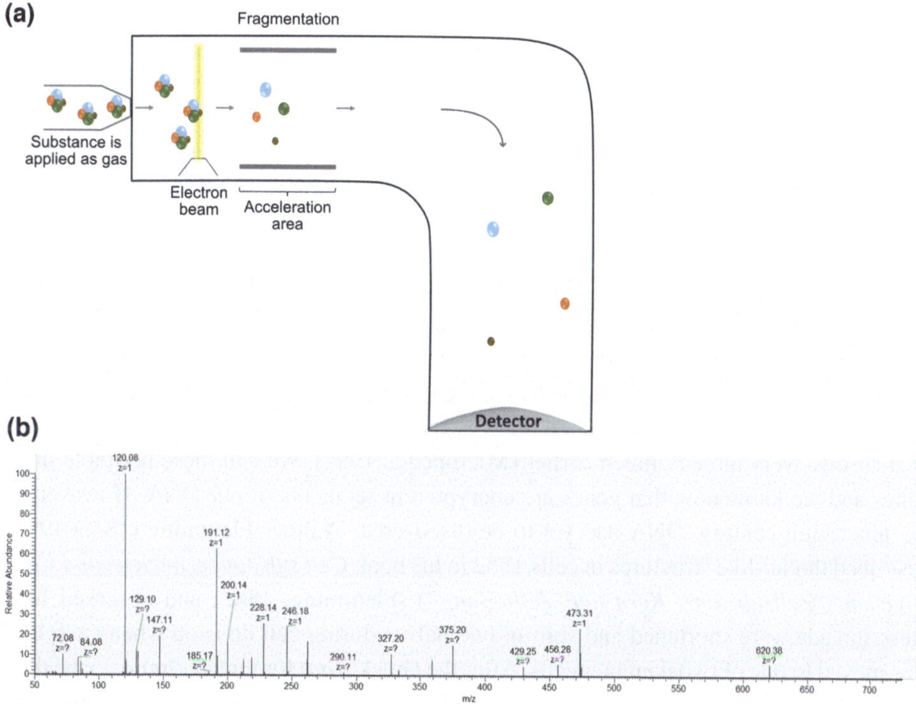

Fig. 1.29 Mass spectrometry; (**a**) Schematic showing mass spectrometer with sample in the gas phase being ionized and accelerated, deflection by electromagnetic forces according to mass/charge relationship of sample components, and detector, which they hit at certain positions; (**b**) Output of a mass spectrometric analysis, diagram indicates peaks, *Y*-axis refers to relative abundance of particles in these peaks, *Y*-axis shows "time of flight," expressed as mass/charge ratio (m/z)

1.3.3 Genetics

The obvious resemblance between parents and offspring in all life forms has certainly interested people at all times of history. Aristotle formulated his understanding of inheritance in his book *Generation of Animals*. Evaluating his insights, Devin Henry states in his article *Aristotle on the mechanism of Inheritance*: Aristotle divides the "nature" of a biological substance into its "material nature" and its "formal nature," the latter being then "sub-divided" into what he calls nature "as mover" … and nature as "end." In this, he seems to be drawing roughly the same distinction as modern biology makes between the genotype and the phenotype (Henry, 2006). From these Aristotelian thoughts of the fourth century B.C., it is a long leap into the nineteenth century when Gregor Mendel performed his first experiments in a small monastery garden in the Augustinian monastery of Brünn, which he had entered in 1843. He crossed varieties of garden peas displaying different tallness, flower colors, and others, and observed how they were maintained through the generations. His experiments resulted in his description of now-called Mendelian laws, which explain the principles of heredity of characters. He reported them in 1965 to the "Brünn society for the study of natural sciences" when he stated himself that "among all the numerous experiments made, not one has been carried out to such an extent and in such a way as to make it possible to determine the number of different forms under which the offspring of hybrids appear, or to arrange these forms with certainty according to their separate generations, or definitely to ascertain their statistical relations." For instance, he showed that crossing red and white flowered peas resulted in 100% of pink offspring in the first generation. However, in the next generation, red and white plants were segregated again at a constant ratio of 25% white, 25% red, and 50% pink. He concluded that there must be pairs of elementary units of heredity, and the offspring always inherited one half of the pair from each parent (Fig. 1.30). Similar results were obtained in 1900 by Carl Erich Correns, Erich Tschermak von Seysenegg, and Hugo de Fries independently; however, they realized and acknowledged that Gregor Mendel had published the major points of their discovery already much earlier (Micropedia, 1992). We call these heritable units genes and we know now that genes are encrypted in segments of our DNA. However, in the nineteenth century, DNA had yet to be discovered. Walther Flemming (1843–1905) described thread-like structures in cells 1882 in his book *Cell substance, nucleus and Cell-Division* (*"Zellsubstanz, Kern und Zellteilung"*) (Flemming, 1882) and observed how these threads were shortened and split in two halves during cell division when each half was moved to one of two daughter cells. After the Greek word for thread ("mitos"), he then coined the term mitosis for the process of cell division (Micropedia, 1992). Today, his threads are called chromosomes. From these discoveries, eventually, it was recognized that Mendel's "pairs of elementary units of heredity," which are the basis of Mendelian

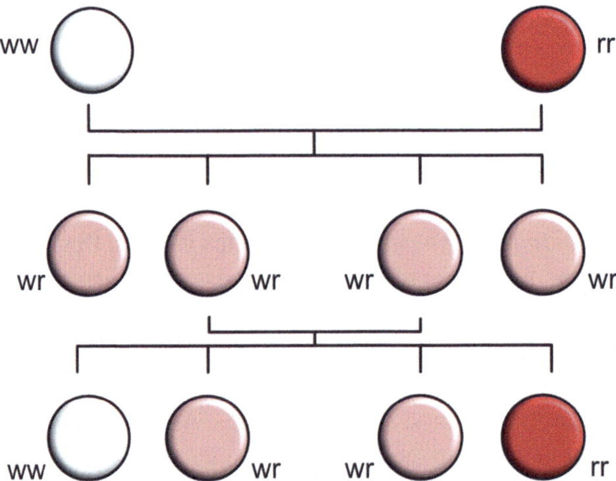

Fig. 1.30 Mendelian heredity: If each parent has two identical versions of hereditary units (now called alleles) for a certain trait, e.g., red or white flower color, all offspring will be the same because it inherits one version of this unit from each parent. However, the unit versions (alleles) are split in the next generation between the offspring, here 25% white, 25% red, and 50% mixed

laws of heredity, were physically localized on chromosomes. From 1907 to 1920, Thomas Hunt Morgan studied the fruit fly *Drosophila melanogaster* and found morphological mutants, which he could map to single chromosomes (of which *Drosophila* only has 4). This was the dawn of genetic experiments with model organisms, where *Drosophila melanogaster* is still on top of the list, even if flies are now accompanied by other genetically engineered animals like the nematode *Caenorhabditis elegans*, the fish *Danio rerio,* and the mammal *Mus musculus*. They all produce offspring in large numbers with short generation times, have naturally occurring mutants, which can be analyzed and are amenable to experimental mutagenesis. Together with the toolbox of molecular biology allowing plasmid DNA construction, transfer of DNA into cells and organisms and DNA sequencing, the investigation of mutant strains of these model organisms has proven a very powerful tool in life science research. More often than once, biochemical and genetic lines of research have suddenly converged and new pathways were discovered. This will be explained in more detail when growth factor research and apoptosis are discussed (Chap. 5).

It was a long way to go before DNA had been acknowledged as the carrier of our hereditary information. In 1914, Robert Feulgen developed a staining procedure for nucleic acids using hydrochloric acid treatment followed by Schiff's reagent (fuchsinschweflige Säure) staining and resulting in red to violet color accumulating in the nucleus. However, this was not taken as an indication that the aforementioned "threads" could contain nucleic acids. DNA was in fact considered too "primitive" a molecule to be responsible for carrying the great amount of information necessary for heritage.

In 1928, Frederick Griffith showed the first genetic transformation when he was working with strains of *Pneumococcus*, a bacterium responsible for lung disease. We all have *Pneumococcus* bacteria quietly sitting in our epithelia, but sometimes, they cause an outbreak of sinusitis or even pneumonia. Griffith distinguished a strain with a smooth surface embedded in a mucous capsule (S) and one with a rough surface not protected by a capsule (R). When mice were infected with S, they died of Sepsis, and with R, they remained healthy. He then used the harmless R strain in combination with dead S strain bacteria inactivated by heat, or with cell-free extracts of S. In both cases, the mice died, showing that there was a transmittable (or transforming) molecule in the S strain. This was followed by 16 years of intensive biochemistry to identify this molecule. It was eventually shown in 1944 by Oswald Avery, Colin MacLeod, and Maclyn McCarty that the transforming molecule was present in the nucleic acid fraction of the bacterial lysates and that it could be destroyed by DNAse, an enzyme that degrades DNA, therefore the transforming molecule was declared to be DNA. The next hint at DNA as a transforming agent came from phage research. Phages were first described in 1917 by the Canadian Hubert d'Herelle and the Georgian Georgi Eliavaare. They found that the sewage from hospitals contained an invisible agent that infected bacteria and when agar plates fully grown with bacteria were treated with it, plaques became visible where bacteria had been lyzed. Therefore, they were called phages ("bacteria eaters"). They are practically viruses that infect bacteria. Georgi Eliavaare initiated a whole phage bank in Georgia where phages are collected and separated on the basis of their host bacteria. We can distinguish *Coli* phages, *Salmonella* phages, *Staphylococcus* phages, and many more. Today, it is estimated that 10^{30} virions are present in the sea water. Phages infecting pathogenic bacteria present a big hope for treatment of antibiotic-resistant bacterial strains. Phages contain a head structure filled with DNA with a collar sitting on an enveloped tube where DNA can be passed along. It terminates in the endplate, which attaches to bacterial surfaces with small fibers (Fig. 1.31a). Alfred Hershey and Martha Chase, by differentially labeling phage protein with a ^{35}S-radioactive isotope and nucleic acid with a ^{32}P -radioactive isotope, found that it was nucleic acid (DNA), which was responsible for lyzing bacteria (Hershey & Chase, 1952), Fig. 1.31b. In 1952, Erwin Chargraff analyzed the chemical composition of DNA and detected Adenine, Guanine, Thymine, and Cytosine bases, where always Adenine/Thymine and Guanine/Cytosine occurred at the same amounts (Chargaff et al., 1952). In 1953, Watson and Crick presented the double helix structure of DNA (Franklin & Gosling, 1953; Watson & Crick, 1953), followed by the decoding of the information that was presented as triplets of nucleotides in the DNA sequence. They were awarded the Nobel Prize in 1962.

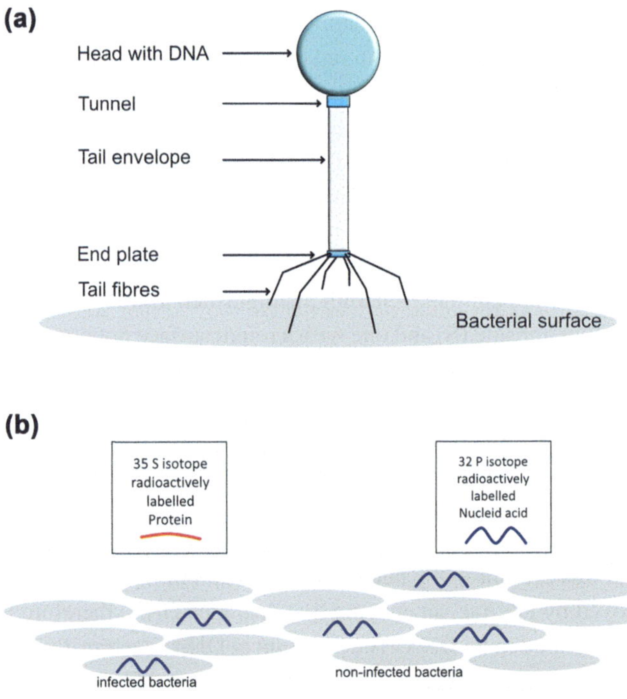

Fig. 1.31 (**a**) Structure of a bacteriophage with DNA-filled head, tunnel and tail structure, end plate and tail fibers; (**b**) Gray area indicates bacterial surface where phage attaches and "injects" DNA into the bacterium; (**b**) Hershey and Chase experiment using phages radioactively labeled for either protein (^{35}S) or DNA (^{32}P) to infect bacteria, after phage replication bacterial fractions of infected bacteria were only labeled with ^{32}P, indicating that DNA and not protein had been introduced into bacteria during infection and was responsible for bacterial lysis

Their insights were very much based on structural analyses of Rosalind Franklin; however, she died in 1958, too early to receive the 1962 Nobel Prize together with Watson and Crick.

Eventually, DNA was taken seriously as a carrier of hereditary information. Nevertheless, already in 1943, in a series of talks entitled "Was ist Leben?" (What is Life?), Erwin Schrödinger had formulated a vision of what genes could be by saying: "Genes represented aperiodic crystals which, by a kind of morse alphabet message, would be regulating the development of the cell."(*"Gene wären aperiodische Kristalle, die in einer Art "Morsebotschaft" die Entwicklung der Zelle steuern."*) For these genes carried by chromosomes, he envisaged properties exceeding simple carriage of information, as they should "contain in some kind of code script the entire pattern of the future development and of its functioning in the mature state" (Schrödinger, 1943).

In April 2022, in the journal "Science," the complete sequence of a human genome with 23 million bases was published (Nurk et al., 2022). Previous sequencing efforts of the human genome had always been short of ca. 8% of the whole sequence. Have we now decoded the essence of life? Sometimes, it seems as if humankind was at the verge of succumbing to the idea of a complete understanding of all physical aspects of life and therefore embracing the possibility to control them. We do know a lot, we can genetically modify plants, and we even dare to put viral genes into our bodies, but remember that we cannot explain how a single molecule of H_2O can adapt such different forms as liquid, steam, snowflakes, and ice crystals! We should be modest and appreciate the insights we are given by acknowledging that the power of natural life is probably beyond our current imagination.

Test Yourself

1. How big are cells?
2. What are the differences between pro- and eukaryotic cells?
3. What are cell organelles?
4. What is the largest cell organelle?
5. How can cell organelles be separated biochemically?
6. How can proteins be purified and analyzed?
7. How has genetics contributed to cell biological research?

References

Balavoine, G., & Adoutte, A. (1998). One or three cambrian radiations. *Science, 280*, 397–398.

Böttger, V., Böttger, A., Lane, E. B., & Spruce, B. A. (1995). Comprehensive epitope analysis of monoclonal anti-proenkephalin antibodies using phage display libraries and synthetic peptides: Revelation of antibody fine specificities caused by somatic mutations in the variable region genes. *Journal of Molecular Biology, 247*(5), 932–946.

Cagniard-Latour. (1838). Abhandlung über die weinige Gärung (transl from French). *Annales de Chimie et de Physique, 68*, 206ff.

Chargaff, E., Lipshitz, R., & Green, C. (1952). Composition of the desoxypentose nucleic acids of four genera of sea-urchin. *The Journal of Biological Chemistry, 195*(1), 155–160.

Delbrück, M., & Schrohe, A. (1904). *Hefe, Gärung, Fäulnis*. Verlagsbuchhandlung Berlin Paul Parey.

Denk, W., & Horstmann, H. (2004). Serial block-face scanning electron microscopy to reconstruct three-dimensional tissue nanostructure. *PLoS Biology, 2*(11), e329.

Flemming, W. (1882). *Zellsubstanz, Kern und Zellteilung*.

Folkes. (1723). Some account of Mr.Leeuwenhoek's curious microscopes, lately presented to the Royal Society. *Royal Society's Philosophical Transactions, 32*(380), 446–453.

Franklin, R. E., & Gosling, R. G. (1953). Evidence for 2-chain helix in crystalline structure of sodium deoxyribonucleate. *Nature, 172*(4369), 156–157.

Gerhart, L. M., & Ward, J. K. (2010). Plant responses to low [CO2] of the past. *The New Phytologist, 188*(3), 674–695.

Gomaa, E. Z. (2020). Human gut microbiota/microbiome in health and diseases: A review. *Antonie Van Leeuwenhoek, 113*(12), 2019–2040.

Henry, D. (2006). Aristotle on the mechanism of inheritance. *Journal of the History of Biology, 39*, 425–455.

Hershey, A. D., & Chase, M. (1952). Independent functions of viral protein and nucleic acid in growth of bacteriophage. *The Journal of General Physiology, 36*(1), 39–56.

Hillenkamp, F., & Karas, M. (1990). Mass spectrometry of peptides and proteins by matrix-assisted ultraviolet laser desorption/ionization. *Methods in Enzymology, 193*, 280–295.

Ho, C. S., Lam, C. W., Chan, M. H., Cheung, R. C., Law, L. K., Lit, L. C., Ng, K. F., Suen, M. W., & Tai, H. L. (2003). Electrospray ionisation mass spectrometry: Principles and clinical applications. *Clinical Biochemist Reviews, 24*(1), 3–12.

Hooke, R. (1665). *Micrographia, or some physiological descriptions of minute bodies made by magnifying glasses, with observations and inquiries thereupon*, London.

Kützing, F. T. (1837). *Mikroskopische Untersuchungen über die Hefe und Essigmutter, nebst mehreren naderen dazu gehörigen vegetabilischen Gebilden*. Versammlung des naturhistorischen Vereins des Harzes.

Lindstrom, K., & Mousavi, S. A. (2020). Effectiveness of nitrogen fixation in rhizobia. *Microbial Biotechnology, 13*(5), 1314–1335.

McGuffee, S. R., & Elcock, A. H. (2010). Diffusion, crowding & protein stability in a dynamic molecular model of the bacterial cytoplasm. *PLoS Computational Biology, 6*(3), e1000694.

Micropedia. (1992). *Enzyclopedia Britannica* (Vol. 2, pp. 559–560). University of Chicago.

Miescher, F. (1871). Medizinisch-Chemische Untersuchungen. *HOPPE-SEYLER, 4*, 441–460.

Müller-Taubenberger, A., Vos, M. J., Böttger, A., Lasi, M., Lai, F. P., Fischer, M., & Rottner, K. (2006). Monomeric red fluorescent protein variants used for imaging studies in different species. *European Journal of Cell Biology, 85*, 1119.

Nurk, S., Koren, S., Rhie, A., Rautiainen, M., Bzikadze, A. V., Mikheenko, A., Vollger, M. R., Altemose, N., Uralsky, L., Gershman, A., Aganezov, S., Hoyt, S. J., Diekhans, M., Logsdon, G. A., Alonge, M., Antonarakis, S. E., Borchers, M., Bouffard, G. G., Brooks, S. Y., Caldas, G. V., Chen, N. C., Cheng, H., Chin, C. S., Chow, W., de Lima, L. G., Dishuck, P. C., Durbin, R., Dvorkina, T., Fiddes, I. T., Formenti, G., Fulton, R. S., Fungtammasan, A., Garrison, E., Grady, P. G. S., Graves-Lindsay, T. A., Hall, I. M., Hansen, N. F., Hartley, G. A., Haukness, M., Howe, K., Hunkapiller, M. W., Jain, C., Jain, M., Jarvis, E. D., Kerpedjiev, P., Kirsche, M., Kolmogorov, M., Korlach, J., Kremitzki, M., Li, H., Maduro, V. V., Marschall, T., McCartney, A. M., McDaniel, J., Miller, D. E., Mullikin, J. C., Myers, E. W., Olson, N. D., Paten, B., Peluso, P., Pevzner, P. A., Porubsky, D., Potapova, T., Rogaev, E. I., Rosenfeld, J. A., Salzberg, S. L.,

Schneider, V. A., Sedlazeck, F. J., Shafin, K., Shew, C. J., Shumate, A., Sims, Y., Smit, A. F. A., Soto, D. C., Sovic, I., Storer, J. M., Streets, A., Sullivan, B. A., Thibaud-Nissen, F., Torrance, J., Wagner, J., Walenz, B. P., Wenger, A., Wood, J. M. D., Xiao, C., Yan, S. M., Young, A. C., Zarate, S., Surti, U., McCoy, R. C., Dennis, M. Y., Alexandrov, I. A., Gerton, J. L., O'Neill, R. J., Timp, W., Zook, J. M., Schatz, M. C., Eichler, E. E., Miga, K. H., & Phillippy, A. M. (2022). The complete sequence of a human genome. *Science, 376*(6588), 44–53.

Pfeifer, K., Ergal, I., Koller, M., Basen, M., Schuster, B., & Rittmann, S. K. R. (2021). Archaea biotechnology. *Biotechnology Advances, 47*, 107668.

Remak, R. (1852). Ueber extracellulare Etnstehung thierischer Zellen und über die Vermehrung derselben durch Theilung. *Archiv für Anatomie, Physiologie und Wissenschaftliche Medicin (Müller's Archiv), 19*, 47–57.

Ribatti, D. (2018). An historical note on the cell theory. *Experimental Cell Research, 364*(1), 1–4.

Schermelleh, L., Ferrand, A., Huser, T., Eggeling, C., Sauer, M., Biehlmaier, O., & Drummen, G. P. C. (2019). Super-resolution microscopy demystified. *Nature Cell Biology, 21*(1), 72–84.

Schrödinger, E. (1943). *Was ist Leben?*

Schwann, T. (1939). *Mikroskopische Untersuchungen über die Übereinstimmung in der Struktur und dem Wachstum von Tieren und Pflanzen*. Verlag der Sander'schen Buchhandlung Berlin.

Virchow, R. (1858). *Die cellular pathology*. Hirschwald.

Watson, J. D., & Crick, F. H. (1953). Molecular structure of nucleic acids; a structure for deoxyribose nucleic acid. *Nature, 171*(4356), 737–738.

Biochemistry of the Macromolecules of Life: Nucleic Acids, Proteins, Carbohydrates, and Lipids

2

What You Will Learn in This Chapter

Despite the wide morphological diversity, the main contents of all cells of living organisms are biological macromolecules, including proteins, nucleic acids, lipids, and carbohydrates. These macromolecules are the basis of all cellular functions and forms. Nucleic acids, proteins, and carbohydrates are produced by linking monomers of similar type by a common chemical bond. Nucleic acids are made from nucleotide building blocks joined together by phosphodiester bonds, proteins are made by joining together amino acids by peptide bonds, and carbohydrates are made from monomeric sugar molecules joined by glycosidic bonds. Lipids make an exception. They consist of glycerol, forming ester bonds with fatty acids on each of its three hydroxyl groups. In this chapter, the chemical nature of the nucleic acids DNA and RNA, and of proteins are discussed. In addition, the processes of DNA synthesis by replication in cells and by *PCR* in test tubes, the process of RNA synthesis in cells by transcription and of protein synthesis by translation are explained. Methods of genetic engineering are depicted on this basis. Finally, the chemical nature of carbohydrates and lipids is discussed.

2.1 Nucleic Acids

Nucleic acids include *d*esoxyribo*n*ucleic *a*cid (DNA) and *r*ibo*n*ucleic *a*cid (RNA). DNA is present in the cell nucleus and contains the information for the amino acid sequence of all proteins that are produced in any particular organism. In the form of DNA, this information is inherited by daughter cells after cell divisions. Specific segments of the DNA encode individual proteins. All cell types of our body have many proteins in common, for instance those, which perform the enzymatic reactions necessary for metabolic activity.

A. Böttger, J. Moneer, *Basic Cell Biology for Dentists and Medical Students*, https://doi.org/10.1007/978-3-032-03636-0_2

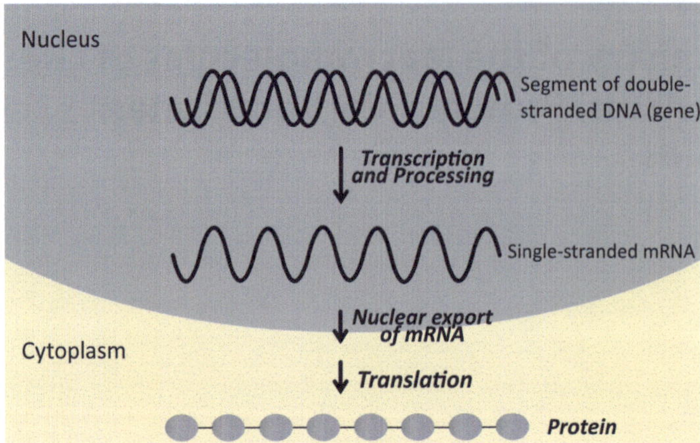

Fig. 2.1 Schematic representation of double-stranded DNA, single stranded mRNA after transcription and processing in the nucleus, nuclear export, and protein after translation in the cytoplasm of the cell

However, all tissues and the cells they are composed of, in a multicellular organism, perform different functions and accordingly are equipped with different proteins. A red blood cell, for instance, contains hemoglobin, whereas a muscle cell is filled with actin and myosin proteins. Therefore, in a well-controlled manner, only segments of DNA, also called genes, which encode the required proteins, are activated in any given cell. These are then transcribed into pre-messenger *RNA*, which is processed into mature messenger RNA (mRNA) (Fig. 2.1). mRNA contains the same nucleotide code as its DNA template. It is exported out of the nucleus into the cytoplasm, where it is associated with ribosomes. Ribosomes are protein factories, where amino acids are joined together according to the mRNA code, embedded in the sequence of the respective mRNA. This process of manufacturing proteins is called translation (Fig. 2.1). Amino acid chains, after having been knotted by the ribosome, are folded into functional 3D structures with the help of chaperones. After being synthesized and folded, proteins experience many post-translational modifications, including phosphorylation, glycosylation, hydroxylation, and others. Only then, they are functional.

2.1.1 Biochemical Nature of DNA and RNA

Desoxyribonucleic acid (DNA) and ribonucleic acid (RNA) are polymers of nucleotides. They are found in the nucleus, hence the name “nucleic acids”. DNA always resides in the nucleus, RNA, in contrast, is produced in the nucleus, but many species of RNA function in the cytoplasm. These include structural RNAs that are part of cytoplasmic particles such as ribosomes or signal recognition particles; and mRNAs providing the matrix for protein synthesis during translation.

DNA is a double helix, which is formed by Watson–Crick base pairing of nucleotides on either helical strand. RNA is a single strand of nucleotides. Both, DNA and RNA are densely packed together with specific proteins. In the case of DNA, about half of the packaging proteins are histones. These are small proteins with many basic amino acids. DNA has an acidic phosphate backbone, as we will see later, and thus, the positively charged basic amino acids of histones can easily bind to the negatively charged phosphate backbone of the DNA by ionic interactions.

The nucleus can be stained with certain basic dyes, such as hematoxylin, which is used for anatomic preparations. The first DNA dye was developed in 1914 by Robert Feulgen. He treated the tissue first with hydrochloric acid and then added an acidic solution of fuchsin with sodium disulfite, also known as "Schiff's base." This resulted in a dark violet staining of nuclei. Due to the property of DNA to be accessible for color staining, the nuclear content was called "chromatin" (chrome = color). This term is still used. It now refers to DNA that is bounded by proteins. In a diploid human cell, we find 46 independent - meaning not physically connected - DNA double helices representing our 46 human chromosomes. If they were totally unfolded strings, those strings together would have a length of two meters. Hence, two meters of DNA have to fit into a nucleus with a radius of only about 5 μm. This explains why the DNA has to be dramatically compacted. Condensation is achieved by winding the DNA around beads made of histones. There are four different histone proteins, H2A, H2B, H3, and H4. H2A and H2B dimerize and form a tetramer with another H2A–H2B dimer. H3 and H4 do the same. Both tetramers then form an octamer (Fig. 2.2a). The beads, called nucleosomes, thus consist of histone octamers and have a diameter of ca. 11 nm, allowing 147 nucleotide pairs to wind around each one of them. Between two nucleosomes, DNA strings of ca. 80 nucleotides are present, this part is called linker DNA (Fig. 2.2b). Another histone, histone H1, binds the DNA between entering and exiting the nucleosomes and is thereby involved in chromatin compaction (Song et al., 2014). The whole appearance of nuclear chromatin at its lowest condensation level would thus be like beads on a string. However, to fit into the nucleus, nucleosomes are further condensed into fibers forming loops. Folding of those loops compacts the DNA even more and at its highest condensation level, during cell division, chromatin displays visible chromosomes (Fig. 2.2c). During cell division, chromosomes can be easily partitioned into daughter cells. After mitosis, these chromosomes decondense, but the individual strings are not wildly mixed like Mikado-game sticks, each string rather covers a defined territory within the nucleus that is separated from the territory of other strings. This is illustrated in Fig. 2.3. Here, by staining individual strings with fluorescent dyes, researchers made two chromosomes visible, chromosome 1 and chromosome 40. They are seen by fluorescence microscopy in their chromosome state during mitosis, and in their decondensed state after mitosis (so-called interphase, see Chap. 5) (Fig. 2.3) (Cremer et al., 2000).

RNA is divided into several classes and performs many different functions in a cell. *Messenger RNA (mRNA)* is transcribed in the nucleus and then transported out of the nucleus into the cytoplasm where it provides the matrix for protein synthesis. During its formation and nuclear export, mRNAs are tightly bound by a large number of proteins,

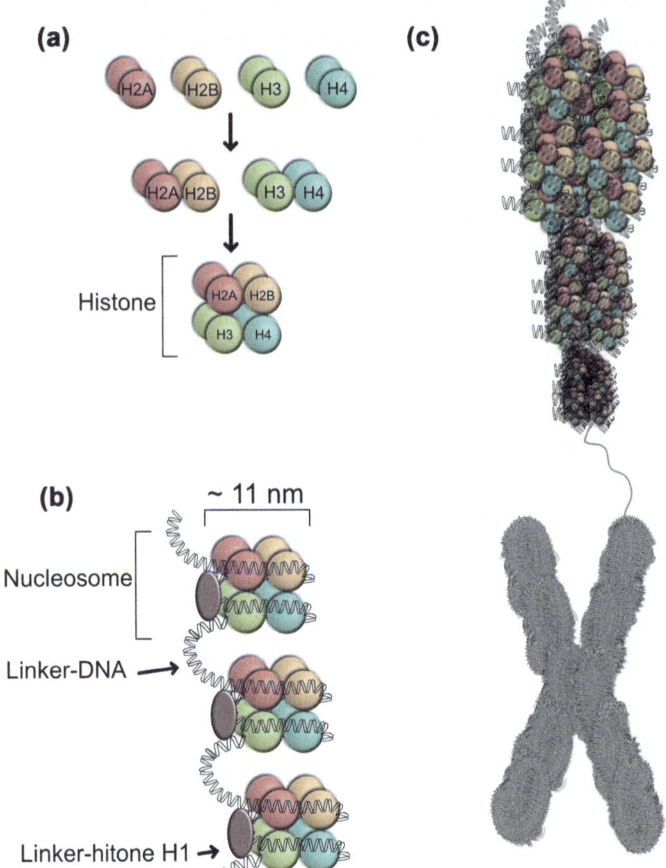

Fig. 2.2 (**a**) Schematic representation of histone octamer building from histone dimers (H2A, H2B, H3, and H4, red, light brown, green, and cyan circles); (**b**) Nucleosomes constituting DNA wound around histone octamers; linker histone H1 (dark brown oval) binding to the nucleosome; (**c**) Further condensation of nucleosomes leading into chromosomes

which provide regulatory elements to control mRNA stability, processing, and the whole process of translation. Transfer RNAs (tRNAs) transfer amino acids to the ribosomes for translation (Fig. 2.4). Moreover, RNAs are structural elements of reaction compartments of cells, for instance, of ribosomes. Ribosomal subunits are composed of proteins and large ribosomal RNAs (rRNAs). The large ribosomal subunit contains 34 proteins (L1–L34), and the small ribosomal subunit contains 21 proteins, S1–S21. These proteins build a complex with RNA, whereby RNA forms a scaffold structure. Forming of a stable 3D-structure from RNA is possible through complementary sequences that are present in different regions of the RNA and thus able to establish Watson–Crick base pairs amongst each other.

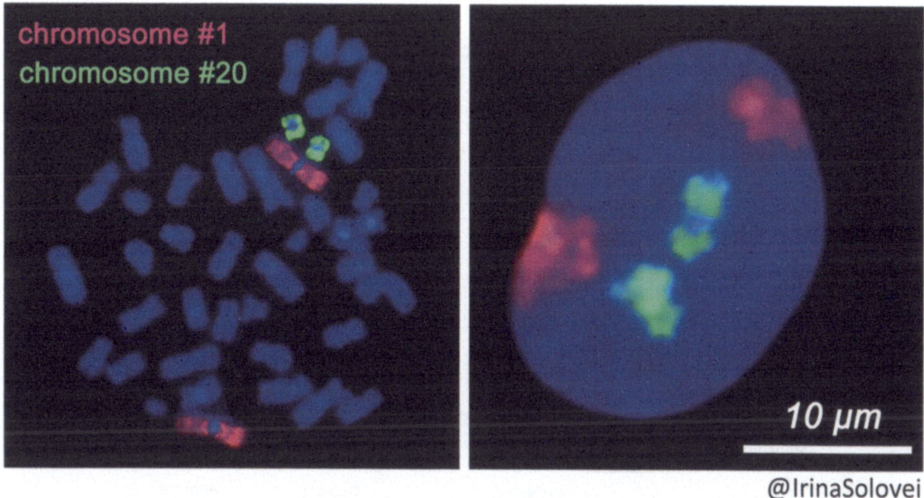

Fig. 2.3 Left-hand panel shows chromosomes in mitotic prophase made visible with DNA stain DAPI, blue; chromosome 1, magenta and chromosome 20, green, are fluorescently labeled; Right-hand panel shows interphase chromatin, chromosome 1 and chromosome 20 are now decondensed and form two "chromosomal territories" each

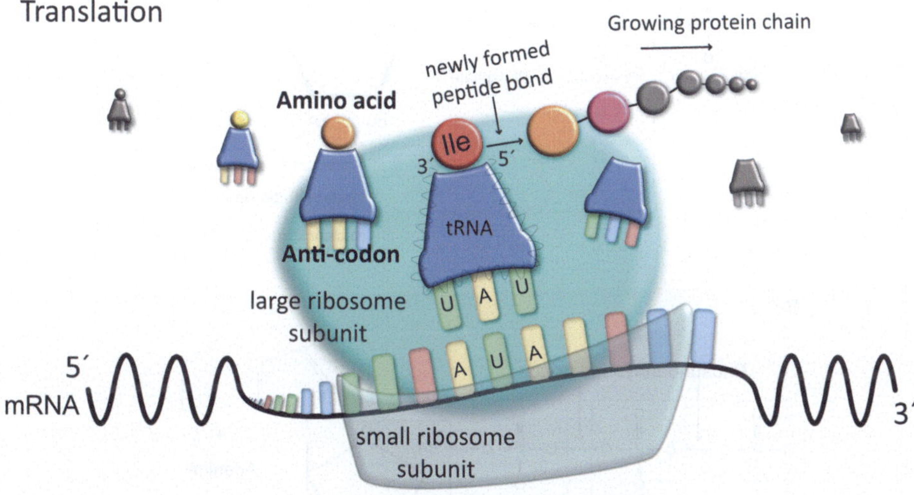

Fig. 2.4 Schematic representation of translation: mRNA positioned between large and small ribosome subunits (gray, light green), coding sequence (AUA) pairs with anticodon on t-RNA; t-RNA is connected to amino acid (Ser), and amino acid (Ser) is then released from t-RNA and added to growing protein chain through a newly formed peptide bond

In recent years, in addition to the above described RNA-species, several regulatory RNAs were discovered, which take part in gene regulation, for instance, microRNAs and small interfering RNAs (siRNAs).

DNA and RNA are polynucleotides, e.g., chains of nucleotides joined by nucleotide bonds (Figs. 2.5 and 2.6). A nucleotide consists of a pentose sugar (ribose or deoxyribose),

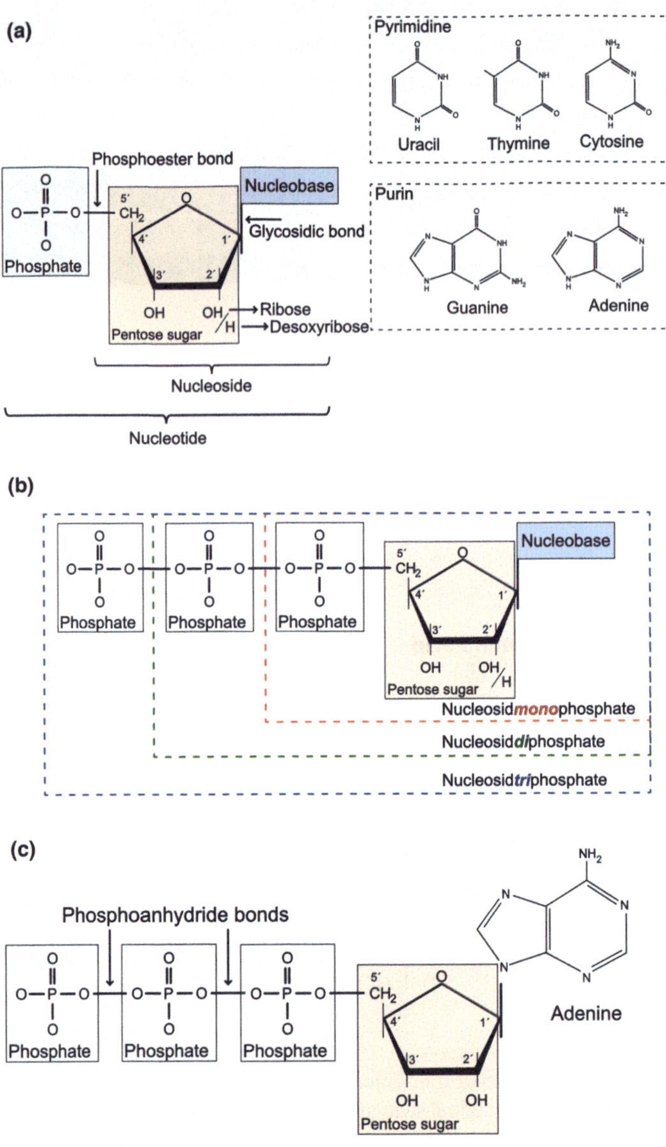

Fig. 2.5 (**a**) General composition of nucleosides and nucleotides, pyrimidine (uracil, thymine, cytosine) and purin nucleobases (guanine, adenine) are shown with structural formulas, ribose has OH in 2′-position of ribose sugar, desoxyribose has H in this position; (**b**) General structure of nucleoside-mono, di-, and triphosphates; (**c**) Chemical structure of Adenosintriphosphate, ATP

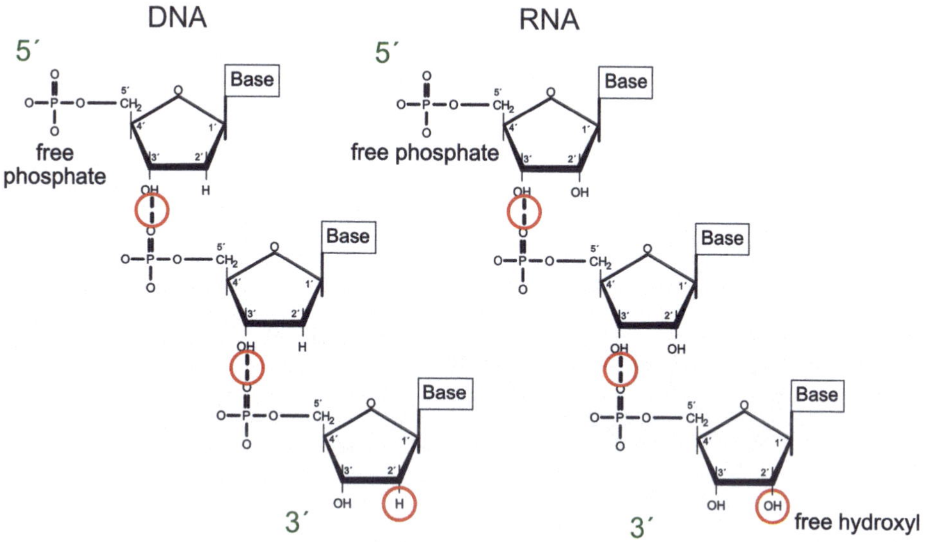

Fig. 2.6 Formation of nucleotide bonds between 3′-OH of one nucleotide and 5′-phosphate of the next one to produce DNA and RNA, 2′H in DNA and 2′OH in RNA are indicated by red circle, 5′ and 3′ ends are indicated by green letters

a phosphate, and a nucleobase. The carbon atoms of the sugar are numbered from 1′ to 5′. A phosphoester bond is formed between a phosphate and the hydroxyl group of the 5′ carbon of the sugar, and a glycosidic bond is formed between the nucleobase and the 1′ carbon of the sugar (Fig. 2.5a). The nucleobases are represented by adenine (A), guanine (G), cytosine (C), thymine (T), and uracil (U). All of these are heterocyclic aromatic compounds (containing ring structures with C and N atoms), the so-called pyrimidine bases (C, T, and U) have one aromatic ring structure, and the two purine bases (A and G) have two (Fig. 2.5a). A nucleobase bound to nonphosphorylated ribose or deoxyribose constitutes a nucleoside. Nucleotides are thus phosphorylated nucleosides and can exist as nucleoside monophosphates, -diphosphates or -triphosphates (Fig. 2.5b). As an example for the adenine base, we thus have adenosine (adenine + ribose), adenosine monophosphate, diphosphate or triphosphate, the latter abbreviated as ATP (Fig. 2.5c). The bond between the ribose and the phosphate is a phosphoester bond; however, the bonds between the phosphates are energy-rich phosphoanhydride bonds.

There are two important differences between RNA and DNA. First, as the name implies, DNA contains the deoxyribose sugar moieties, and RNA contains ribose. Second, DNA contains thymine bases in addition to adenine, guanine, and cytosine (T, A, G, and C), whereas RNA contains uracil bases in addition to adenine, guanine, and cytosine (U, A, G,

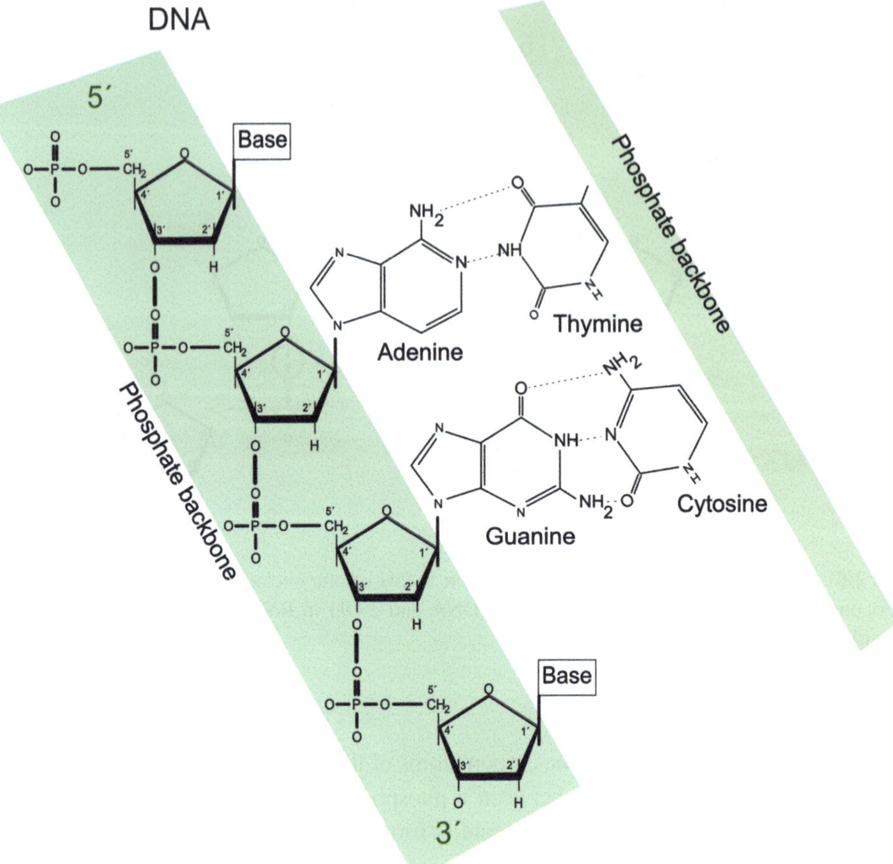

Fig. 2.7 Part of DNA double helix, Watson–Crick base pairing between adenine and thymine and guanine and cytosine are indicated by structural formulas, and phosphate backbone is indicated by green box

and C). In both cases, nucleotides are joined together by nucleotide bonds. These are formed between the 3′-hydroxyl of the deoxyribose or ribose of one nucleotide, and the 5′-phosphate of the next nucleotide. Then, deoxyribose as part of the nucleic acid chain does not have any free hydroxyl groups left. In contrast, RNA retains a free hydroxyl at the 2′-carbon. This enables RNA to use this hydroxyl group for catalytic reactions. DNA does not have any catalytic capacity (Fig. 2.6).

The previous description explains that both DNA and RNA have a phosphate backbone. As had been worked out by Rosalind Franklin and Watson and Crick in 1953 (Watson & Crick, 1953), the structure of DNA is a double helix. Two strands of nucleotide chains are intertwined. The phosphate backbone is directed to the outside and the bases are arranged inside (Fig. 2.7). Two bases from opposite strands pair with each other and form hydrogen bridges. Due to the different sizes of purins (two rings) and pyrimidines (one ring), they can pair. However, the pairing is more defined. Adenine always pairs with thymine, and guanine always pairs with cytosine (Fig. 2.7). This so-called "Watson–Crick base pairing"

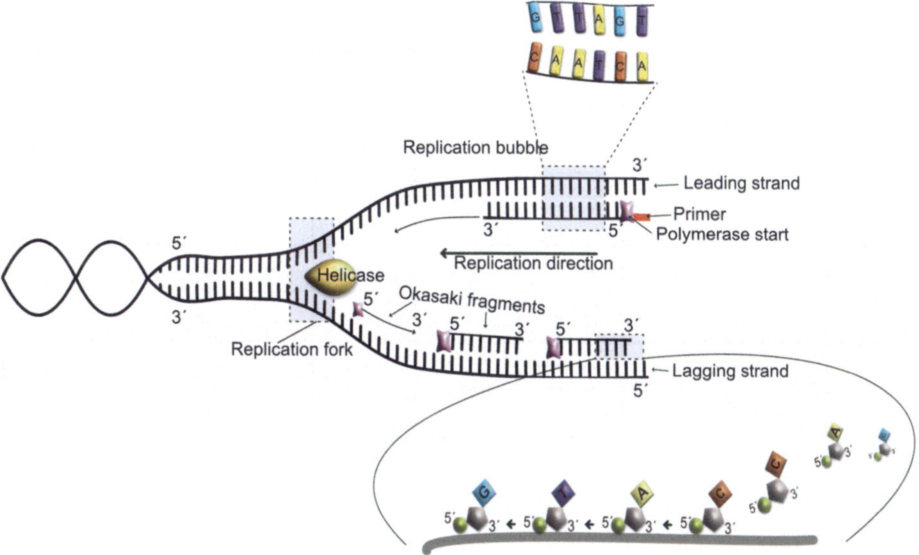

Fig. 2.8 Schematic representation of the DNA-replication process, DNA unwinding by helices opens replication bubble at replication fork, leading and lagging strands and replication direction 5′ to 3′ are indicated, primer is indicated in red on leading strand only; on lagging strand, Okasaki fragments are synthesized by DNA polymerase, nucleotides are always added to 3′end, GTAC are indicated by colors blue, pink, yellow, and orange, respectively

is the reason for sequence conservation of newly synthesized DNA strands during the process of DNA replication (Fig. 2.8). When the two strands are separated, new nucleotides pair with the bases of the mother strand, always A with T and G with C. As synthesis always occurs from 5′ to 3′, the two strands of a DNA molecule are arranged in an antiparallel way. The newly synthesized strand is not identical with the mother strand, but complementary. However, the sequence information remains completely conserved (Fig. 2.8). Doubling of the DNA always occurs before a cell divides. Both daughter cells emerge from the division process with identical DNA sequences. It is the DNA that carries our heritage from generation to generation.

2.1.2 The Genetic Code

Three consecutive nucleotides provide the code for one amino acid. This makes 64 possible codes. Although we only have 20 amino acids and a STOP signal to be encoded, all possible codes are used. Therefore, some amino acids have up to six possible nucleotide codes, and some only have one. In addition, we have three stop codons. Therefore, the genetic code is "degenerated" (Fig. 2.9). In this way, some mutations in the DNA of a protein-encoding gene remain "silent", because a nucleotide exchange does not always lead to an exchange of the respective amino acid.

(a)

	U		C		A		G	
U	UUU	*Phe*	UCU	*Ser*	UAU	*Tyr*	UGU	*Cys*
	UUC		UCC		UAC		UGC	
	UUA	*Leu*	UCA		UAA	*Stop*	UGA	*Stop*
	UUG		UCG		UAG		UGG	*Trp*
C	CUU	*Leu*	CCU	*Pro*	CAU	*His*	CGU	*Arg*
	CUC		CCC		CAC		CGC	
	CUA		CCA		CAA	*Gln*	CGA	
	CUG		CCG		CAG		CGG	
A	AUU	*Ile*	ACU	*Thr*	AAU	*Asn*	AGU	*Ser*
	AUC		ACC		AAC		AGC	
	AUA		ACA		AAA	*Lys*	AGA	*Arg*
	AUG	*Met*	ACG		AAG		AGG	
G	GUU	*Val*	GCU	*Ala*	GAU	*Asp*	GGU	*Gly*
	GUC		GCC		GAC		GGC	
	GUA		GCA		GAA	*Glu*	GGA	
	GUG		GCG		GAG		GGG	

(b)

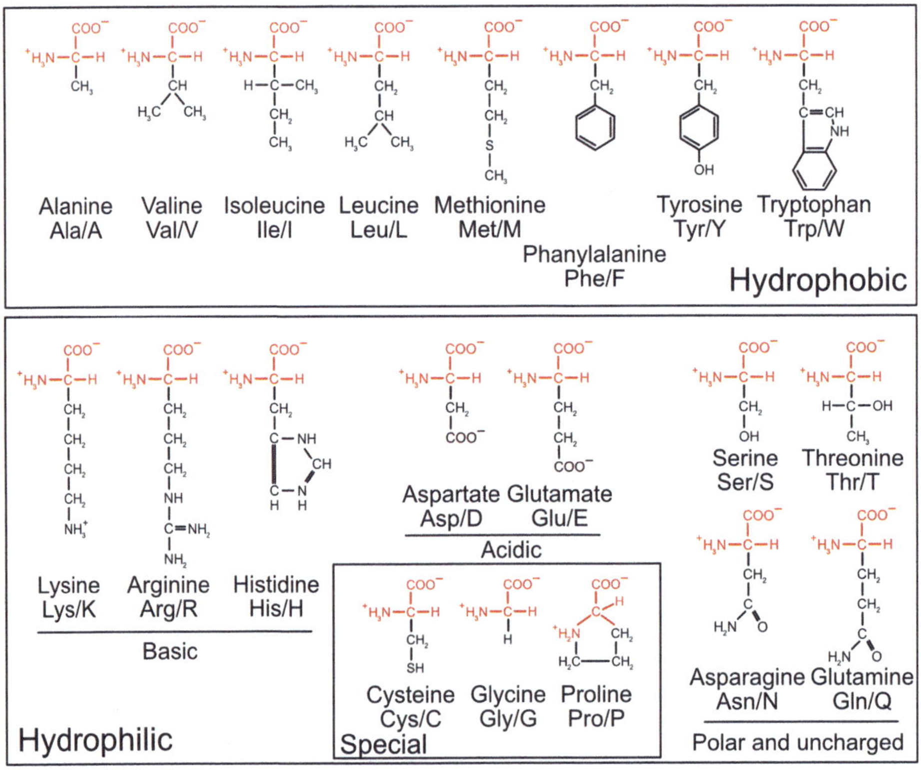

Alanine Ala/A; Valine Val/V; Isoleucine Ile/I; Leucine Leu/L; Methionine Met/M; Phanylalanine Phe/F; Tyrosine Tyr/Y; Tryptophan Trp/W

Hydrophobic

Lysine Lys/K; Arginine Arg/R; Histidine His/H — Basic

Aspartate Asp/D; Glutamate Glu/E — Acidic

Cysteine Cys/C; Glycine Gly/G; Proline Pro/P — Special

Serine Ser/S; Threonine Thr/T; Asparagine Asn/N; Glutamine Gln/Q — Polar and uncharged

Hydrophilic

Fig. 2.9 (**a**) Genetic code; (**b**) Amino acids, structural formulas of residual R sorted according to their chemical nature (hydrophobic, hydrophilic including acidic and basic, polar and nonpolar amino acids, the amino acids cysteine with SH group, glycine as the smallest amino acid with only one H attached instead of a residual R, and proline with ring formation are summarized as special, full names are given in addition to three letter codes and one letter symbols; (**c**) table summarizing amino acids with names and symbols listed according to charge (positive, negative, and uncharged) and polarity (polar, nonpolar)

(c)

Asp	negative		Ala	nonpolar
Glu	negative		Gly	nonpolar
Arg	positive		Val	nonpolar
Lys	positive		Leu	nonpolar
His	positive		Ile	nonpolar
Asn	uncharged polar		Pro	nonpolar
Gln	uncharged polar		Phe	nonpolar
Ser	uncharged polar		Met	nonpolar
Thr	uncharged polar		Trp	nonpolar
Tyr	uncharged polar		Cys	nonpolar

Fig. 2.9 (continued)

2.1.3 DNA Is Synthesized During Replication

Replication starts at different positions on each chromosome at the same time. These positions are called "origins of replication." Early in the division cycle of a cell, specific proteins are arranged at these positions to prepare for the replication process, which starts when a cell enters the cell cycle and starts the "synthesis" phase (S phase). The two DNA strands are then separated creating a "replication bubble" with two fork-sites that move into the unseparated part of the DNA molecule as replication continues. DNA helicases are required for opening of the DNA at these sites. DNA is synthesized by adding nucleotides with their 5′-phosphate to the 3′-OH-end of a pre-existing DNA strand. This means that a DNA strand cannot be synthesized without a pre-existing nucleic acid, where nucleotides can be added on. DNA polymerase needs a primer to start its work. These primers in the cell are short pieces of RNA. They are produced by RNA polymerases. A specialist enzyme for production of primers for replication is primase, an RNA polymerase that produces short-lived RNA oligonucleotides only for use by DNA polymerase during replication. These RNA nucleotides base pair with DNA at the replication bubble and allow nucleotides to be added at their 3′ end. Note that RNA has a uracil base instead of a thymidine, and the uridine of the RNA primer will base pair with adenine. The newly synthesized strand that moves into the replication fork toward the 5′end of the old DNA strand can be synthesized without interruptions, and this is the leading strand. The strand on the other side of the replication bubble, the lagging strand following behind the leading strand, cannot be synthesized continuously because with every opening of the replication fork, a primer has to be added. This is due to the given direction of nucleotides joined together from 5′ to 3′ (adding on at the 3′ end). Therefore, only fragments are synthesized. These are called "Okasaki fragments" and they will be eventually joined together by a DNA ligase. This is illustrated in Fig. 2.8.

DNA replication in a dividing cell is restricted to one round. Then, the cell has two copies of its DNA which can be distributed equally to the two daughter cells during mitosis. Some cells even in the human body are polyploid, e.g., cells in our liver. This state occurs when DNA is duplicated without cell division.

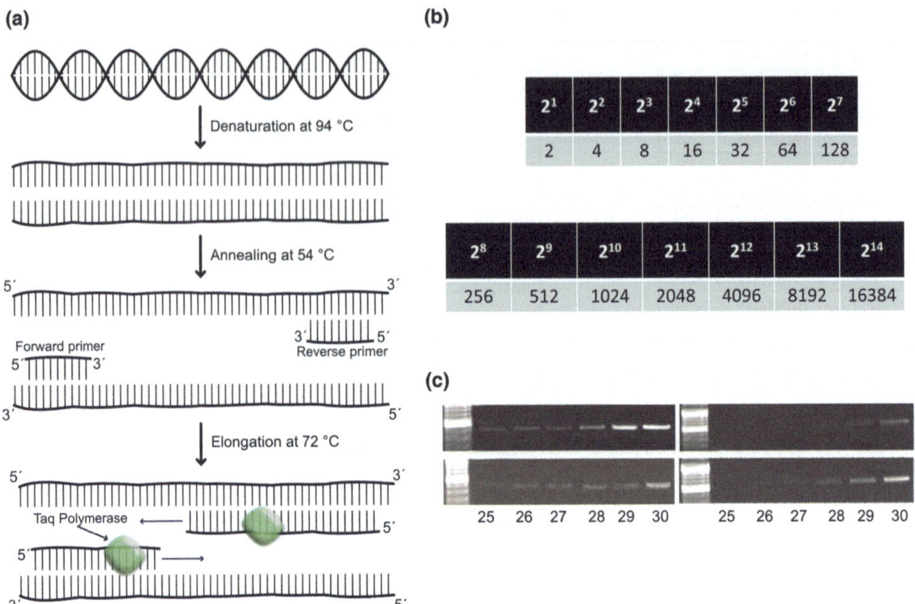

Fig. 2.10 (**a**) Schematic representation of PCR steps including denaturation, and annealing of primers and elongation of PCR products are indicated, Taq-polymerase is represented in green; (**b**) number of PCR product molecules accumulating at cycles 2–14; (**c**) gel electrophoresis of PCR products obtained after indicated numbers of cycles, PCR products are visible under UV due to Ethidium bromide-staining of DNA

2.1.4 DNA Synthesis by Polymerase Chain Reaction (PCR)

If a replication cycle is repeated again and again, DNA can be amplified exponentially. This does not happen in a cell; however, the scientist Kary B. Mullis developed a method to amplify DNA in a test tube, the polymerase chain reaction, in short, *PCR*. By using a DNA-template and pairs of DNA primers that anneal 5′ and 3′ with these specific DNA-sequences, then adding DNA-polymerase and nucleotides to the mixture and carrying out repeated cycles of separating the DNA strands, one can obtain millions of copies of the original piece of DNA (Fig. 2.10a). Kary B. Mullis described the way this idea occurred to him very entertainingly in his autobiographical book *Dancing Naked in the Mind Field* (Mullis, 2000). For his PCR method to work efficiently, he was lucky to be able to take advantage of a very heat-stable DNA polymerase, the Taq polymerase. This polymerase had been isolated from bacteria of *Thermus aquaticus* living at temperatures of 70–75 °C, which were discovered in 1969 by Brook and Freeze in hot springs in the YellowStone National Park in California (Brock & Freeze, 1969). PCR requires temperature cycles which include the melting (separation of DNA strands) temperature for the whole DNA double strand at 95 °C. By using this heat-stable DNA polymerase, the PCR process could easily be automated. For his invention, Kary B. Mullis received the Nobel Prize in 1993.

PCR is a very sensitive method. Within only a few minutes one PCR cycle, during which the DNA is doubled, is completed, with every new cycle the amount of PCR product doubles (Fig. 2.10b). From a minimal amount of a DNA sample, e.g., forensic material, one can amplify as many copies as needed for detection and characterization.

The PCR product is a piece of DNA, which classically is detected by running it through an agarose gel electrophoresis and staining it with some dye that intercalates with DNA, such as ethidium bromide (Fig. 2.10c). The size of the PCR product will be estimated by comparison with a DNA ladder marker, a mixture of DNA fragments with defined sizes (as shown in Fig. 1.27). More recent methods follow the emergence of the PCR product directly. These are called "real-time" PCR methods. They measure fluorescence that arises in one of two ways during the synthesis of the PCR products. In one case, a fluorophore is added to the PCR. It intercalates with double-stranded DNA being produced cycle by cycle of the PCR. This intercalation results in a conformational change of the fluorophore whereby it becomes able to absorb blue light and emit green fluorescence. As the PCR proceeds, more and more dsDNA will be produced and thus, the fluorescence signal increases in proportion with the amount of newly synthesized DNA (Fig. 2.11a). In the second case, in addition to the two primers, a probe is used. This will hybridize with the suspected PCR product DNA somewhere in the middle between the two primers. It possesses a fluorophore and a molecule that quenches the fluorescence of the fluorophore. During the PCR, the polymerase passes the probe and degrades it. When this happens, the quencher is separated from the fluorophore and the latter emits fluorescence (Fig. 2.11b). The output of a real-time PCR is a diagram that shows the fluorescence after each cycle. After a lag phase, where no fluorescence is measured, an exponential increase of the signal is seen, which then goes into saturation when all the nucleotides are used or the polymerase activity expires. The diagram in Fig. 2.11c shows the amount of PCR product after each cycle as it appears in real time. The amount of amplified DNA is related to a *Ct*-value (cycle *t*hreshold), which indicates the number of cycles needed for the fluorescence signal to pass beyond a certain threshold. The more DNA was present before PCR amplification, the less cycles are needed to pass the threshold. Thus, low Ct-values indicate a high amount of sample DNA, and high Ct-values indicate a low amount of sample DNA (Fig. 2.11c). The forward and reverse primers are designed in such a way that they bind specifically to the DNA fragment of interest. However, more often than never, primers can also bind to the DNA in a less specific way. Thus, nonspecific DNA fragments can be amplified and after enough cycles, this results in nonspecific output. In order to develop a stable and informative PCR assay, which allows quantitative measurements of certain DNA specimens, thorough investigations have to be performed. This includes establishing a correlation between the Ct-value and the amount of original DNA by normalizing the assay with standard amounts of the target DNA. Moreover, after each reaction, a test has to be performed, which ensures that only one PCR product is amplified. In addition, for standardizing such assays before widespread application in scientific assays or even in diagnostics, the PCR product has to be sequenced. Finally, if the DNA is amplified from mixed materials such as body fluids, and the result is used to find changes in a particular DNA under certain conditions

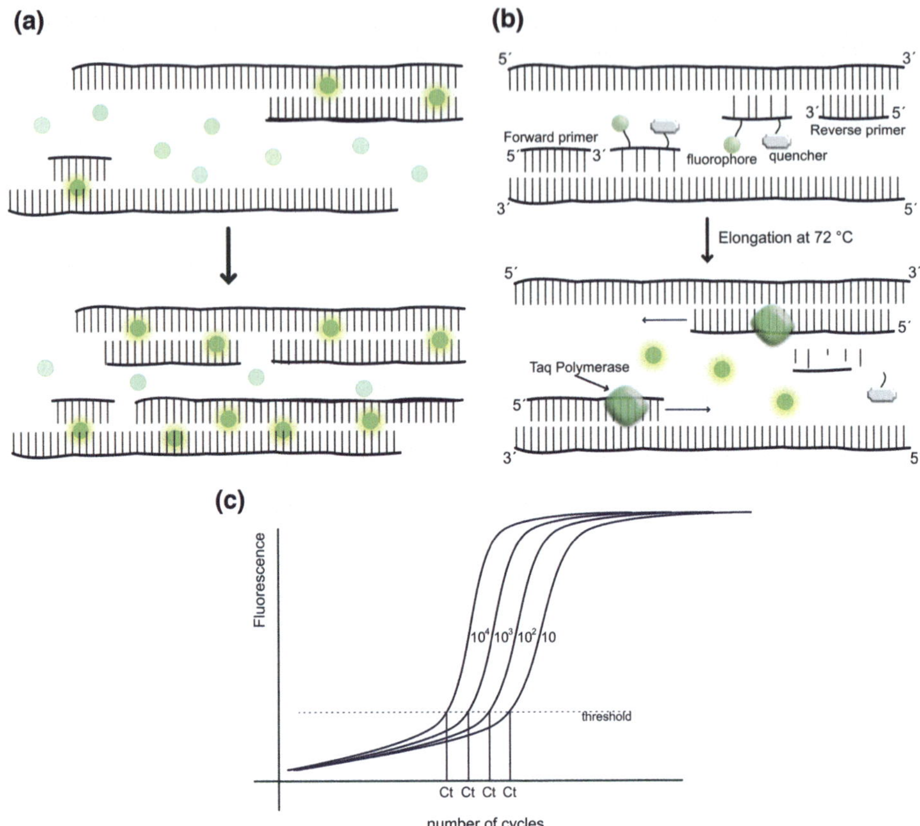

Fig. 2.11 Fluorescence detection for quantitative real-time PCR: (**a**) Case 1: adding fluorophore; fluorophore (light green) fluorescence increases when it intercalates with double-stranded DNA, DNA content doubles with every PCR cycle (shown by green dots); (**b**) Case 2: probe with quenched fluorescence (green circle indicates fluorophore, and gray symbol indicates quencher), destruction of probe by TaqPolymerase releases the fluorophore from the quencher; fluorescence is emitted; (**c**) diagram showing accumulation of fluorescently labeled PCR products during cycles, and Ct values indicate the number of cycles required for reaching a certain threshold of fluorescence; they depend on the amount of DNA present in the sample before amplification begins, for a certain range (in this example, from 10 to 10^4 molecules), the Ct-value is directly proportional to the amount of original DNA

(e.g., infection versus noninfection), Ct-values have to be normalized to nucleic acids in the sample that are not variable under these conditions. As a general rule, it has been established that above the number of 30, Ct-values are often not meaningful. For using PCR tests as a means of diagnosis, they have to be related to symptoms. For example, if a patient has very severe symptoms of an airway infection, it may be useful to detect which virus is causing these symptoms. As there are many different viruses that could be responsible, several of them have to be tested. Anything else would obscure the diagnosis.

Moreover, here, the virus load would have to be directly related to the Ct-values of a diagnostic qPCR, at least for calibration of the PCR test, before it is used on patients. Due to the high sensitivity of the PCR detection method, healthy people might easily produce positive results, especially at Ct-values >30. Here, we refer to the Corman–Drosten PCR test developed for diagnosing SARSCoV2. In the initial description of the test, a direct correlation to nucleic acid of the virus is not provided; moreover, correlation with symptoms is also lacking. The protocol of this PCR test initially used 45 cycles (Corman et al., 2020a, b). To declare patients as infectious on such a basis appears untenable.

PCR requires a double-stranded template DNA. Therefore, it cannot be used on RNA. This is important also for the SARSCoV2 test, because the genetic material of Coronaviruses is RNA (for simplicity, this fact was ignored in the previous explanations). Yet, such PCRs are still possible, when RNA is first retrotranscribed into "copy" DNA (cDNA) by specific enzymes, so-called "reverse transcripases". Reverse transcripases conduct three enzymatic reactions to produce double-stranded cDNA from a single strand of RNA. These include an RNA-dependent DNA polymerase activity, which adds a complementary DNA strand to the RNA template producing a DNA–RNA hybrid, an RNaseH activity (Ribonuclease H), which destroys the RNA strand in the hybrid, and finally, a DNA-dependent DNA polymerase, adding the second DNA strand to the orphaned copy-DNA (Fig. 2.12). Reverse transcriptases were first discovered in retroviruses in 1970 by Howard Temin and David Baltimore (Nobel Prize 1975 with Renato Dulbecco) (Baltimore, 1970). This discovery destroyed the prevailing dogma at the time that the flow of genetic information was unidirectional from DNA via RNA to proteins.

Reverse transcription also happens in human cells. One example is a subunit of telomerase, an enzyme present in stem cells and in some tumor cells. It replaces telomeric repeats, which are lost from chromosome ends in each round of replication. It syntheses new DNA ends on the basis of an RNA-template. Moreover, as will be discussed later, the human genome contains retrotransposons, mobile genetic elements, which can jump from place to place within the genomic DNA, thereby being transcribed into RNA and then retrotranscribed into DNA again to be reintegrated into the DNA at a different position. They encode in their nucleic acid part the proteins necessary for excision, retro-transcription, and insertion.

2.1.5 From DNA to RNA: Transcription

In the cell, RNA is obtained on the basis of the DNA template by the process of transcription. This process requires RNA polymerases. We have three different RNA polymerases in our cells, including RNA polymerase 1 transcribing ribosomal RNA, RNA polymerase 2 transcribing messenger RNA, and polymerase 3 transcribing tRNAs and others. The toxin of the "death cap" fungus, *Amanita phalloides*, α-amanitin, is a potent inhibitor of RNA polymerase 2 and is responsible for the deadly toxic effect of this fungus. When RNA polymerase 2 transcription is blocked, this especially affects cells with high transcriptional activity, such as liver cells, leading to liver failure. The specificity of α-amanitin

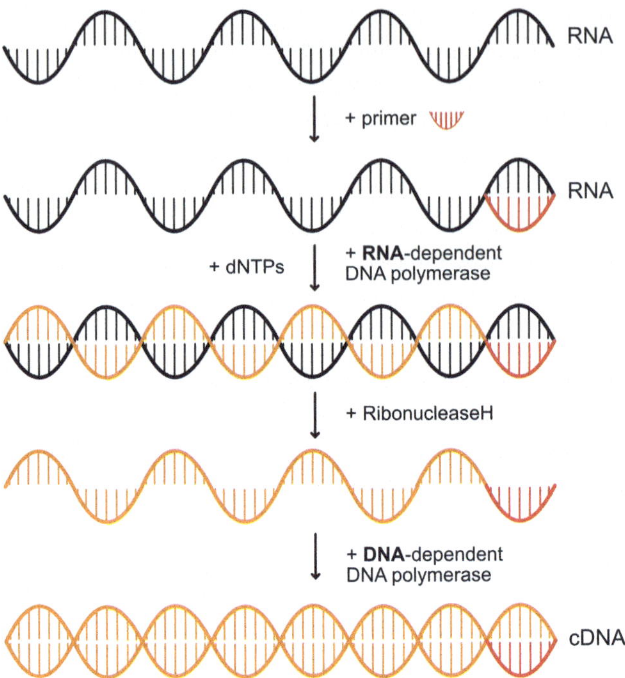

Fig. 2.12 Reverse transcription of RNA templates into cDNA: The enzyme used is *reverse transcriptase* and has three activities; (1) RNA-dependent DNA polymerase synthesizing a DNA strand by base pairing with the RNA template strand; (2) Ribonuclease H degrading the RNA strand and (3) DNA-dependent DNA polymerase synthesizing double-stranded DNA by base pairing with the single-strand DNA

for inhibiting RNA polymerase 2 has been exploited by researchers to study the molecular functions of this enzyme.

Every cell in our body has the same protein-encoding DNA (with some meaningful exceptions); yet, we have many different cells and tissues. RNA polymerase 2, the one that produces messenger RNA, is therefore always recruited to very specific sites in our genome to transcribe the genes that are needed in a particular cell at a particular time of its life cycle. A *gene* is defined as a sequence of our DNA containing information to transcribe RNA. This involves the information of the amino acid sequence of a protein or the nucleotide sequence of an RNA, which have to be produced in this cell at this specific time. In addition, a gene has different parts of sequences to control transcription. These DNA sequences are referred to as cis-regulatory elements.

In the case of mRNA transcription, cis-regulatory elements include a core promoter, to which the RNA polymerase 2 is directly recruited. It is located close to the site where transcription will start. In very tightly regulated genes, upstream of the core promoter, ca.

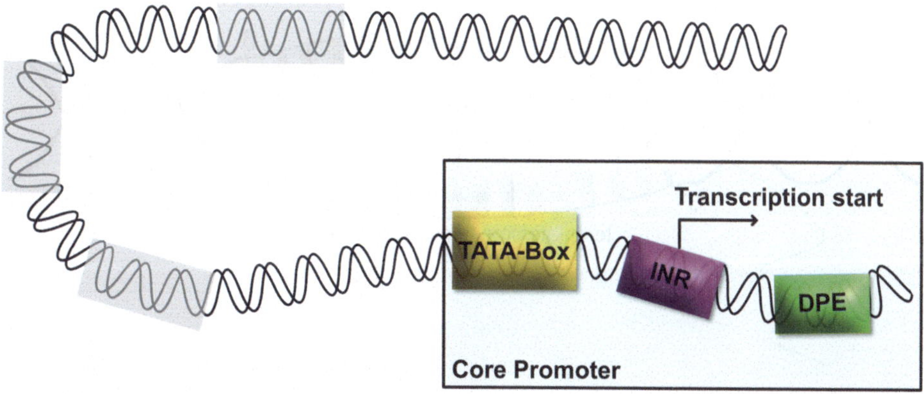

Fig. 2.13 Regulatory region of a gene with core promoter sequences including TATA box (yellow), INR (initiator element, which functions as promoter of transcription in the absence of TATA box (magenta), and DPE (downstream promoter element) (green); Upstream regulatory sequences are indicated as gray boxes

25 nt (nucleotide) from the start site of transcription, there is a TA-rich sequence, the so-called TATA box. Some genes that are less strictly controlled have a "downstream promoter element" (DPE) near the core promoter and constitutively expressed genes (meaning that they are always transcribed in every cell) have an initiator element (INR) (Fig. 2.13). In addition, each gene is regulated by a multitude of protein factors, rarely less than ten and usually more than 30. Thus, further cis-regulatory elements are present within a gene sequence, which can be bound by transcription factors to activate or inhibit transcription of a certain locus on the DNA strand. This regulation happens by either recruiting the RNA polymerase to these sequences (enhancer) or by blockage of RNA polymerase recruitment (silencer). Regulatory elements may be found 5′ of the transcriptional start site, 3′ of the end of the transcript or even in intermittent sequences between the protein-coding regions, so called introns (Fig. 2.13). In this way, the transcriptional information that belongs to a gene may comprise several times the number of nucleotides that are eventually transcribed to mRNA and translated to a protein. In order to investigate a gene, researchers commonly study several thousands of nucleotides upstream of the actual transcriptional start site of a gene, looking for regulatory elements. Identification of these regulatory elements allows the prediction of regulatory processes controlling the respective gene. The transcriptional process is very complex and involves many different proteins. Even the temporal context of the nucleus plays a role in the decision whether transcription of a certain gene occurs or not and to what extent. The regulatory elements represent binding sites for proteins that play a role in the transcriptional event. Which elements become occupied by the corresponding proteins depends on many factors including signals that are transmitted into and out of cells.

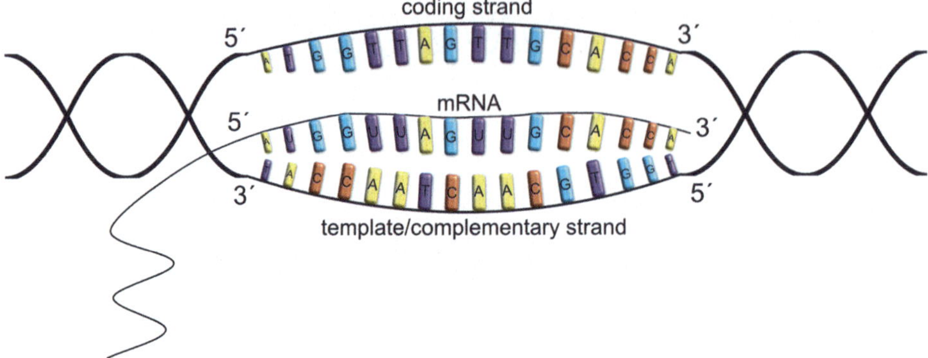

Fig. 2.14 Schematic representation of transcription process, coding, and template strand of DNA are indicated, transcription uses template strand, Pro-mRNA sequence matches sequence of coding strand, and nucleotides ATUGC are illustrated by colored letters

The two DNA strands are distinguished as a coding strand and a template strand. The coding strand provides the open reading frame for translating every triplet of nucleic acids into an amino acid. However, transcription occurs on the opposite strand, the template strand. The RNA polymerase is positioned at the transcription start site of the template strand through interactions with the core transcription factors that themselves are recruited through the activities of specific transcriptional regulators, occupying enhancers. It then joins nucleotides that assemble with this strand according to Watson-Crick base-pairing rules. G pairs with C and A pairs with U ribonucleotides. This obviously results in an emerging RNA sequence, which is identical to the coding strand sequence, except that it contains uracil (U) instead of thymine (T) (Fig. 2.14).

For genes transcribed by RNA polymerase 2, the core promoter is activated by "transcription factor 2" (TFII). TFII represents large protein complexes consisting of several subunits each. In genes with a TATA box, transcription is initiated when the first subunit of the TFII transcription factor complex, the TATA-binding protein (TBP), binds to the TATA box. By binding to TBP, the DNA is bent. This recruits all other subunits of TFII in a stepwise fashion to the core promoter. The last subunit of TFII is TFIIH, which phosphorylates RNA polymerase 2 and unwinds the DNA for transcription to begin (Fig. 2.15a). This transcription results in emergence of a pre-mRNA. It contains parts, which encode protein sequence ("open reading frames") interspersed with parts that do not contain "open reading frames." The former are called exons, the latter, introns. Introns are cut out of the pre-mRNA by the process of mRNA-splicing, resulting in mature mRNA. Moreover, mRNA is equipped with a number of adenine nucleotides (a polyA-tail) at the end and a protective structure at the beginning, the 5' Cap (Fig. 2.15b). The whole way of transcribed RNA from pre-mRNA to mature mRNA is called mRNA processing.

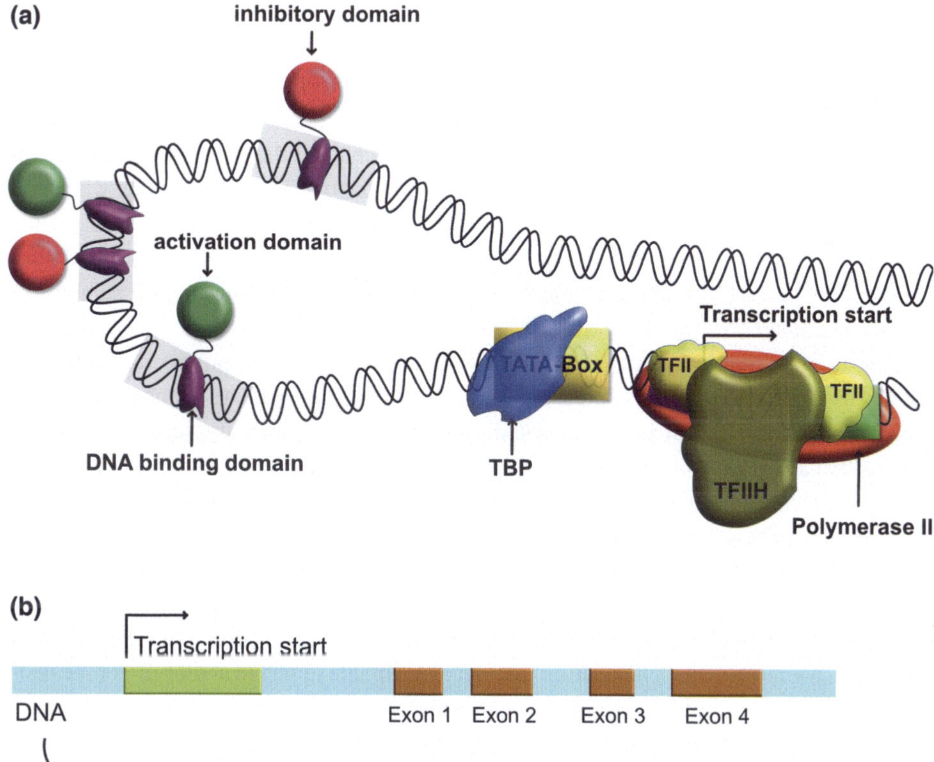

Fig. 2.15 (**a**) Transcription factors binding to regulatory sequences of DNA; A DNA-binding domains specific for certain DNA sequences are shown in pink; activation (green spheres) and repressor (red spheres) domains are indicated, TATA box (yellow box) bound by TATA box-binding protein (TBP, blue) and TFII (yellow form) recruiting RNA polymerase II are indicated at transcriptional start site; (**b**) gene sequence including transcription start, exon and intron-sequences, pre-mRNA with introns and polyA-tail after transcription, and mature mRNA after splicing with polyA-tail and 5′ cap are shown as indicated

The regions of DNA (e.g., genes) to be transcribed are defined by environmental needs (such as the need for nutrition or oxygen), position of the cell in the body, developmental time, and others. This information is transduced to the transcription machinery by a large variety of transcription factors, which themselves are under strict transcriptional control. About 2000 transcription factor genes have been annotated in the human genome.

They usually have a modular structure consisting of a specific DNA-binding domain and an activation (or repressor) domain. DNA-binding domains have a small number of clearly defined tertiary structures that are suited to bind to DNA in a sequence-specific manner. These include homeodomains, Zinc-fingers, basis helix–loop–helix domains and leucine zippers. Often, these proteins form dimers and half of the dimer intercalates with the same binding motif on opposite strands of the DNA. This is possible when the cis-regulatory sequences are palindromic. In this way, each DNA-binding domain fits closely into a certain enhancer or silencing sequence, which is located in the regulatory region of the gene. The function of the activation domains is to recruit the TFII subunits to the core promoter.

Transcription requires that the DNA can be accessed by the RNA polymerase complex. As discussed before, DNA is intensely packed up with the help of histones and other proteins. However, not all DNA is packed up with the same intensity, resulting in DNA regions that are more compact than others. The degree of condensation changes when the histone molecules, around which the DNA is wound, are modified by post-translational modifications. These modifications are reversible and constitute one of several ways in the nucleus to control gene transcription and thus, gene expression. Especially, the ε-amino groups of histone lysine residues can be modified in a number of ways, including methylation (adding one, two or three methyl groups) and ubiquitination (adding the small protein ubiquitin). Moreover, phosphorylation (adding a phosphate to serine and/or threonine residues), or acetylation (adding the acidic acid acyl-residue) (Allfrey et al., 1964; Bannister & Kouzarides, 2011) can occur (Fig. 2.16). Another form of histone modification includes hydroxylation of the C5 carbon atom of lysine (Unoki et al., 2013). As a side note, this hydroxylation depends on oxygen tension, which is also an important parameter for the enzymes that remove methylation marks from histones. This indicates the possibility of a direct impact of cellular oxygen content on histone modification and transcriptional activity in our cells (Schneider & Shilatifard, 2006). Modifications, which add positive charges to the histone tails, such as acetylation, generally promote decondensing of chromatin, whereas nonacetylated histones promote a more condensed state of chromatin (Fig. 2.16). Enzymes which change the acetylation state of histones are thus able to induce a switch between "open" chromatin favoring transcription and "closed" chromatin preventing it. Histone acetyl transferases (HAT) transfer acetyl residues to lysine residues on histone ends and thereby decondense chromatin. Their counterplayers, histone deacetylases (HDAC), remove acetyl residues from histones and favor chromatin condensation. Thus, transcriptional activators often interact with HATs, whereas transcriptional repressors recruit HDACs (Fig. 2.16). Many more histone modifications have been shown to be involved in transcriptional regulation, e.g., proteins that use ATP to push nucleosomes apart. These are chromatin remodelers.

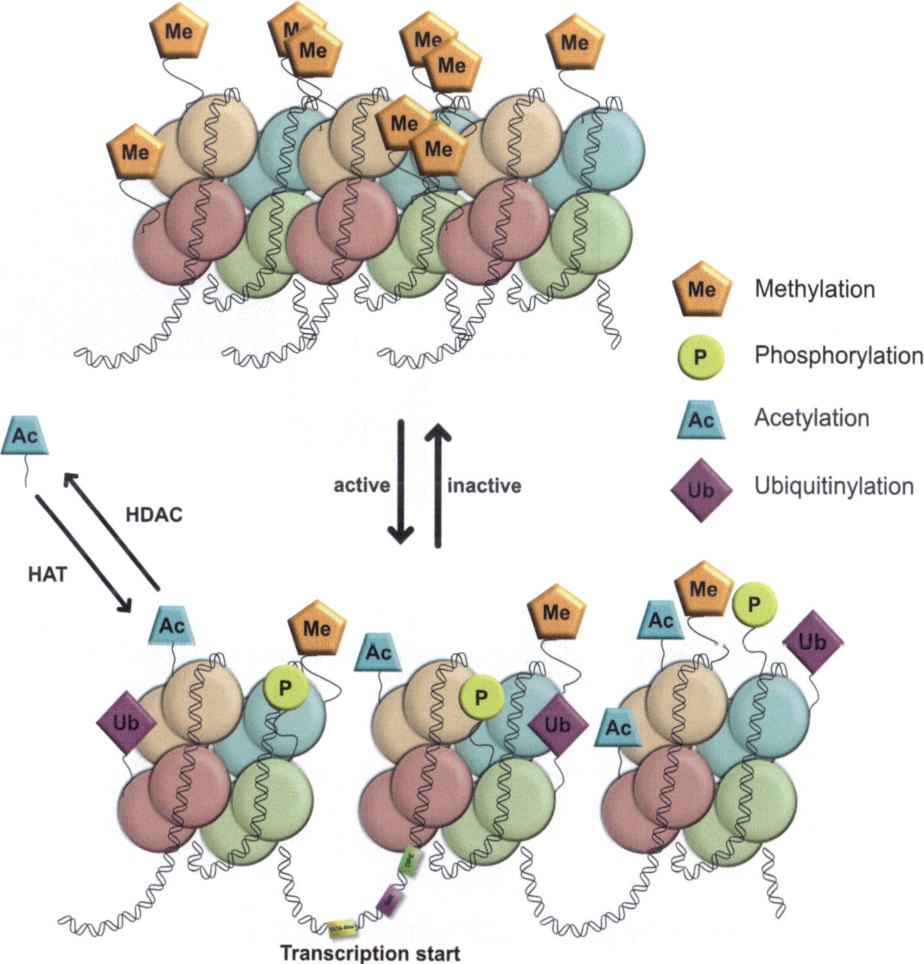

Fig. 2.16 Histone modifications including methylation, phosphorylation, acetylation, and ubiquitinylation of histone ends, inactive chromatin is characterized by absence of acetylation marks, acetylation is carried out by the enzyme histone acetyl transferase (HAT), and deacetylation is carried out by histone deacetylase (HDAC); transcription start sites are "open" in active chromatin state

2.2 Proteins

2.2.1 Proteins Consist of Amino Acids Connected by Peptide Bonds

All proteins are composed of amino acids. An amino acid exhibits one central carbon atom (C1 or Cα) to which four different moieties are attached including a carboxyl group (–COOH), an amino group (–NH_2), and a hydrogen atom (–H). The fourth remaining moiety is a variable side chain, the so-called "residual" R, which is specific for each amino acid (Fig. 2.17 and 2.9b for structural formulars of residualy Rs). There are 20 naturally

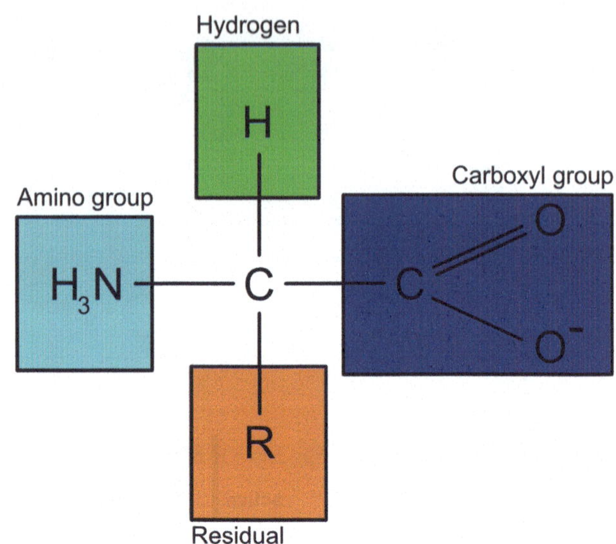

Fig. 2.17 General structural formula of amino acids, central C atom is bonded with amino group (cyan box), hydrogen (green box); carboxyl group (blue box), and a residual structure (R, orange box), for structural formulas of residues R refer to figure 2.9b

Box 2.1

The relative atomic mass of a molecule is given in Dalton (Da). By definition, one Dalton corresponds to 1/12th of the atomic mass of the carbon atom and is also called u = unified atomic mass unit

$$6.022 \times 10^{23} \text{ u have a mass of } 1\text{g}$$

$$1 \text{ mol contains } 6.022 \times 10^{23} \text{ particles } (\text{e.g., molecules})$$
$$(\text{Avogadro - number})$$

Thus the "molecular weight" really is a molecular mass.

$$1 \text{ mol carbon contains } 6.022 \times 10^{23} \text{ molecules and weighs } 12\,\text{g}$$
$$1 \text{ mol water contains } 6.022 \times 10^{23} \text{ molecules of } H_2O \text{ and weighs } 18\,\text{g}$$

occurring amino acids with an average molecular mass of 113 Dalton (Da). Amino acids can be linked together by peptide bonds to form long chains, the so-called "polypeptides". All proteins of life thus consist of variable sequences of 20 different amino acids. The order of these amino acids for each individual protein is encoded in the genome. Proteins comprise either a single polypeptide chain or several polypeptides that form one functional complex. The totality of all so far-identified yeast proteins (called the yeast proteome), for instance, is a collection of about 6600 different protein sequences. The average yeast protein contains 466 amino acids adding to an average molecular mass of ca. 53,000 Da (for explanation of the term molecular mass see Box 2.1). Some proteins are much larger than the given average for yeast proteins, e.g., collagen has a molecular mass of 300,000 Da.

Others are much smaller, for instance, insulin with 6000 Da. Many amino acids needed to produce human proteins are produced in our cells through anabolic pathways. However, the amino acids valine, methionine, leucin, isoleucine, phenylalanine, tryptophan, threonine, and lysine, for children also tyrosine, are so-called essential amino acids, which have to be taken up with our daily diet. Phenylalanine, tyrosine, and tryptophan can only be synthesized in chloroplasts of plants (Böttger et al., 2018).

The properties of each protein entirely depend on the nature and composition of its amino acids. The amino acid sequence of each protein is encoded by a three-nucleotide-code, which is inherited with the DNA, transcribed to mRNA, and used for synthesizing proteins at ribosomes (Fig. 2.9a). Amino acids can be grouped according to their molecular properties (Fig. 2.9b). Lysine and arginine are positively charged, and aspartic acid and glutamic acid are negatively charged. Serine, threonine, and tyrosine have hydroxyl groups, making them accessible for phosphorylation by ester bonding with phosphoric acid residues. They are classified as polar, but uncharged, together with asparagine and glutamine. Leucine, isoleucine, valine, phenylalanine, and others are nonpolar amino acids. Tyrosine and phenylalanine have aromatic side chains. Methionine and cysteine have sulfhydryl groups. The smallest amino acid is glycine. Biochemists employ a three-letter code for each amino acid, as shown in Fig. 2.9a, b. This is easily related to the chemical name of each amino acid. However, writing three letters seems to have become too tedious and was replaced by a one letter code, which is not entirely self-explanatory anymore.

2.2.2 Primary, Secondary, Tertiary, and Quaternary Protein Structure

The *primary structure* of proteins is defined by the sequence of amino acids it is made of. A simple unfolded protein chain exhibits a certain spatial structure due to the defined arrangement of the carbon orbitals. Depending on their identities, the amino acids present in these chains, arrange in specific *secondary structures*. Most prominently, they form α-helices, β-strands, which can be arranged in β-sheets, and single loops (β-turns). These secondary structures are stabilized by hydrogen bonds between the hydrogen of the amino part of the peptide bond and the oxygen of another peptide bond (Fig. 2.18a). α-helixes turn every 3.6 amino acids, thus 10 turns require 36 amino acids and grow to the length of 5.6 nm (Fig. 2.18a).

Different secondary structures can form over the whole length of the amino acid chain. Their spatial arrangement defines the *tertiary structure* of proteins. This can take specific shapes. For instance, long helices can intertwine to form coiled coils (Fig. 2.18b), such as those present in the proteins that make our hair. ß-sheets (Fig. 2.18c) can form barrel-like structures (Fig. 2.18d) as in porins, which are able to establish pores in mitochondrial membranes, and as in the famous "green fluorescent protein" (GFP). Mixtures from β-strands, -turns, and helices can result in compact globular proteins, like actin and tubulin (see later).

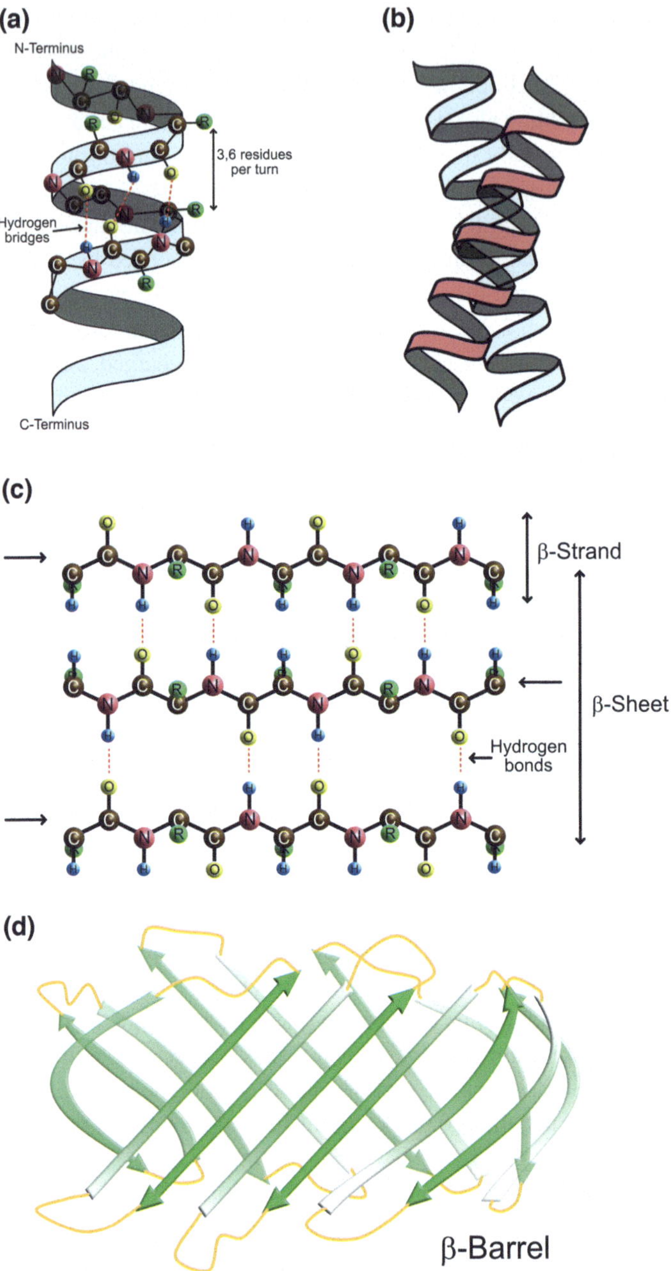

Fig. 2.18 Secondary structure of proteins: (**a**) α-helix with hydrogen bonds; (**b**) two α-helices intertwined; (**c**) β-sheet with β-strands, hydrogen bonds indicated by dotted red lines; (**d**) β-barrel; β-strands are indicated as green arrows with tip pointing toward C-terminus of protein; Loops between β-strands are yellow

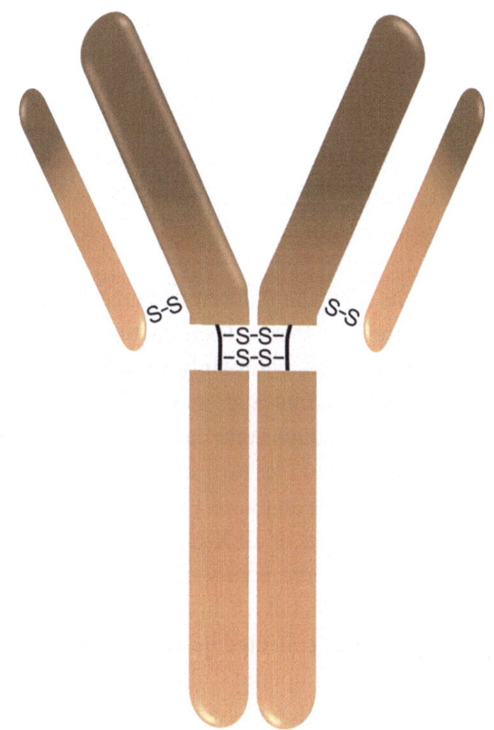

Fig. 2.19 Quaternary structure of a protein, schematic quaternary structure of immunoglobulin (IgG) is shown, two long chains (heavy chains) and two short chains (light chains) represent the IgG structure, and chains are connected by disulfide bonds

Some proteins consist of several subunits. This means that numerous protein chains are assembled together to yield a fully functional protein complex. Such assembly of subunits represents the *quaternary* structure. Immunoglobulins of the IgG-type in humans, for example, consist of two long amino acid chains and two short amino acid chains, resulting in a tetramer. The individual chains are stabilized by disulfide bonds (Fig. 2.19).

Proteins have many functions in the cell, for instance they can form cellular structures, such as the cytoskeleton, they can work enzymatically to catalyze biochemical reactions, or they may function as enzyme regulators. For their purposes, they physically interact with other macromolecules in the cell. Such interactions are defined by the shapes of proteins, which fit into the shape of a binding partner like a key into a lock. In some cases, the shape of one or both partners changes when they interact, and this is called an "induced fit." Protein–protein interactions are usually stabilized by noncovalent bonds, which allow great flexibility in binding and separating of the partners. Noncovalent bonds include ionic bonds between side chains of positively and negatively charged amino acids, hydrogen bonds, where a hydrogen from a hydroxyl or amino group in the side chain may be attracted by an oxygen from a carboxyl group, and hydrophobic interactions, also called "van der Waals" interactions between uncharged side-chain residues of amino acids, for instance, methyl groups. Depending on the number and strength of such noncovalent interactions, the binding affinity of two proteins can be very weak or very strong, or anything in between. Most proteins can bind to multiple interactors including other proteins, peptides, small molecules, polysaccharides, nucleic acids, or lipids.

The primary structures for most proteins can be deduced from the DNA sequences that are available from many genome- and transcriptome- (the totality of all mRNA sequences detected in an organism) sequencing projects. However, from the primary sequence of a protein, its secondary structure elements or its complete tertiary structure cannot be predicted (Brenner, 1999). The structures we know are derived from different lines of biochemical research, but mainly from X-ray structural analyses. When X-rays (or Röntgen rays) are targeted through a crystal, they are diffracted by the grid levels of the crystal. From this diffraction, one can deduce the arrangement of the atoms within the crystal. Only proteins that can be tricked into crystallization thereby adapting defined secondary structures, can be analyzed with this method. Nonstructured parts of proteins will not be visible. The results of all X-ray structural analyses of many different proteins are available to the scientific community on different platforms. Moreover, tools have been developed to predict the structure of an unknown protein by comparing its primary sequence with all X-ray structures collected in databases (e.g., the *P*rotein *H*omology/analog *Y* *R*ecognition *E*ngine V 2.0 of Imperial College London, *PHYRE*). X-ray-resolved protein structures are represented by different graphical means, such as C-α backbone traces, ball and stick, and ribbons or solvent-accessible surface graphics.

2.2.3 The Protein Factory: Translation of Proteins at Ribosomes

The amino acid chain is formed by connecting the carboxyl group of one amino acid with the amino group of the next amino acid forming a peptide bond and releasing one molecule of H_2O. Therefore, every protein has a free amino group at one end (which we call N-terminus) and a free carboxyl group at the other end (which we call C-terminus) (Fig. 2.20).

In the cell, peptide bonds are almost exclusively formed by translation of mRNA into proteins at ribosomes. The mRNA sequence is a blueprint for the amino acid sequence of the protein it encodes. mRNA is transcribed from DNA and processed in the nucleus and then, in its mature form, is translocated from the nucleus into the cytoplasm through nuclear pores. In the cytoplasm, ribosomes attach to the mRNA and provide the scaffold and the enzymes for protein synthesis. Amino acids cannot "read" the nucleotide triplet code of mRNA directly. For translation, each amino acid is linked to a molecule of "transfer RNA" (tRNA), which displays an anticodon that is complementary to the triplet code

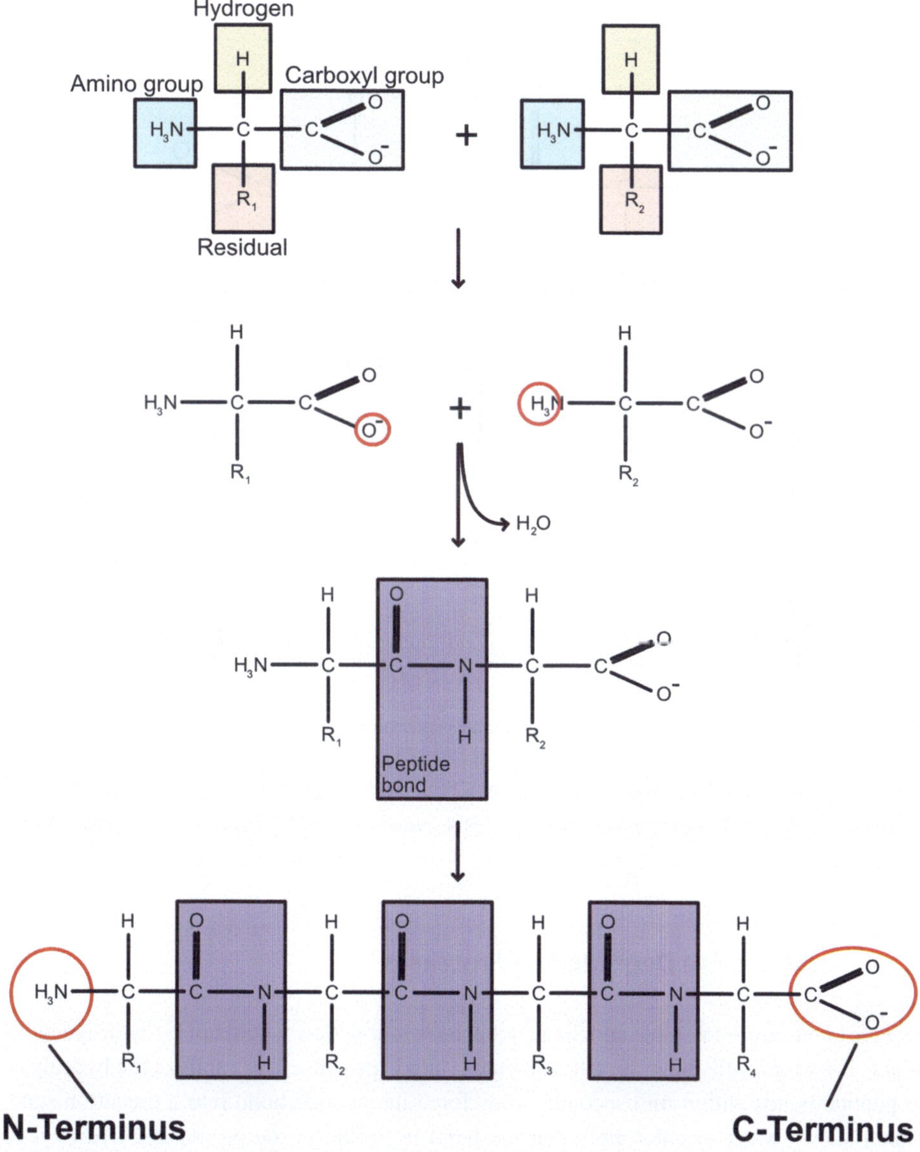

Fig. 2.20 Formation of peptide bond (violet boxes) between carboxyl group and amino group of two amino acids, whereby one molecule of water is released; the direction of peptide bond formation results in free amino group at N-terminus and free carboxyl group at C-terminus

for this specific amino acid. By base pairing with the counterpart of their anticodon on the mRNA strand, t-RNAs carrying amino acids enter the ribosome in the order of the sequence of amino acids required in the protein, which is then synthesized by forming peptide bonds between any two amino acids coming to sit beside each other within the center of the

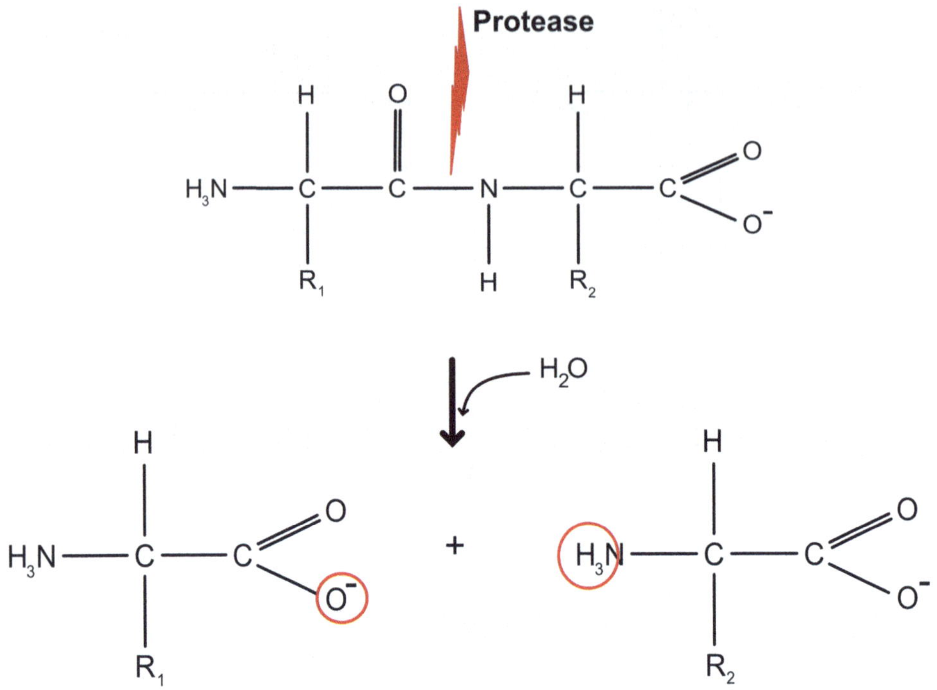

Fig. 2.21 Hydrolysis of peptide bond by proteases (one molecule of water is added)

ribosome. Some t-RNAs base pair with a STOP codon. These are not connected to an amino acid and thus interrupt protein synthesis, causing the polypeptide chain to leave the ribosome (see above, Fig. 2.9a).

2.2.4 Proteins Are Degraded by Proteases

Peptide bonds are very stable and, in an aqueous solution, they would not be hydrolyzed for years. Therefore, cells have specific enzymes called "proteases" to catalyze the hydrolysis of peptide bonds within milliseconds. They force the peptide bond into a transition state, which allows water to enter and open the bond by filling in the proton and hydroxyl to regain free carboxyl- and amino- ends (Fig. 2.21). Proteases are proteins with a certain region in their structure that forms the active site. This region binds the peptides or proteins it wants to cleave (the substrates) on the basis of the sequence of the specific cleavage site, which fits perfectly into the active site of the protease. The catalytic function requires an attack on the carbon atom of the peptide bond using an electron pair. This can be donated by different molecules that are present in the active site (or catalytic center) of the proteases. A large family of proteases exists in cells. They use different catalytic amino acids to attack the peptide bonds. This includes hydroxyl groups in serine- and threonine-proteases, carboxyl groups in aspartate-proteases, sulfhydryl groups as in cysteine-proteases,

Table 2.1 List of some important proteases with examples

Catalytic amino acid or mechanism	Examples
Serine proteases	Coagulation, pancreas, complement, oligopetidases (trypsin, chymotrypsin)
Aspartate proteases	Pepsin, chymosin, presenilin, signal peptidase
Threonine proteases	Proteasom
Metalloproteases	Enkephalinase, prohormone processing enzymes, endopeptidases, matrix metalloproteases
Cysteine proteases	Caspases, cathepsins, calpains.

and also metal ions, as in metalloproteases (Table 2.1). Proteases are found in cells in lysosomes, where they have an acidic pH-optimum. Moreover, they are secreted into body cavities like the stomach or gut, where they digest nutrients. We also know proteases in the blood, such as those making our blood coagulation system. Moreover, cells harbor many large protein complexes called proteasomes, which consist of threonine proteases and degrade specific proteins at certain cellular locations and at certain points of cell physiological pathways. They also degrade damaged or unfolded proteins.

2.2.5 Proteins Are Folded by Chaperones

Of the theoretical number of proteins of average size, based on all possible sequence combinations of 20 different amino acids, only a fraction is found in nature. Not every sequence of amino acids is able to fold into a stable protein. Moreover, amino acid chains as they leave the ribosome after translation are folded into their functional 3D structure. This does not, in most cases, occur spontaneously, but it needs specialized proteins of several protein families, called chaperones. The name was given because a chaperone protein, just as an older aunt chaperoning young girls, may accompany a protein at a certain part of its life cycle, not only during the folding process, but also for preventing arbitrary nonproductive interactions with other proteins in the cell. The name was first suggested for nucleoplasmin, an acidic protein interacting with histones, and described as follows: "We suggest that the role of the protein we have purified is that of a 'molecular chaperone' which prevents incorrect ionic interactions between histones and DNA" (Laskey et al., 1978). Some chaperones, such as nucleoplasmin, only fold one so-called "client" protein, and others are involved in folding several clients. Two large chaperone families are the "heat shock" proteins, Hsp70 and Hsp60. They received this name because their mRNA had been found dramatically upregulated after exposure of cells to high temperatures. The Italian geneticist Feruccio Ritossa was the first to observe striking changes in actively transcribed regions in nuclei of *Drosophila* larvae in 1962. He realized that these were reproducibly induced by exposing larvae to 37 °C. Subsequently, it was shown that the transcribed genes encoded specific protein sequences and thus these proteins were named "heat shock proteins." However, they are not only upregulated by heat shock

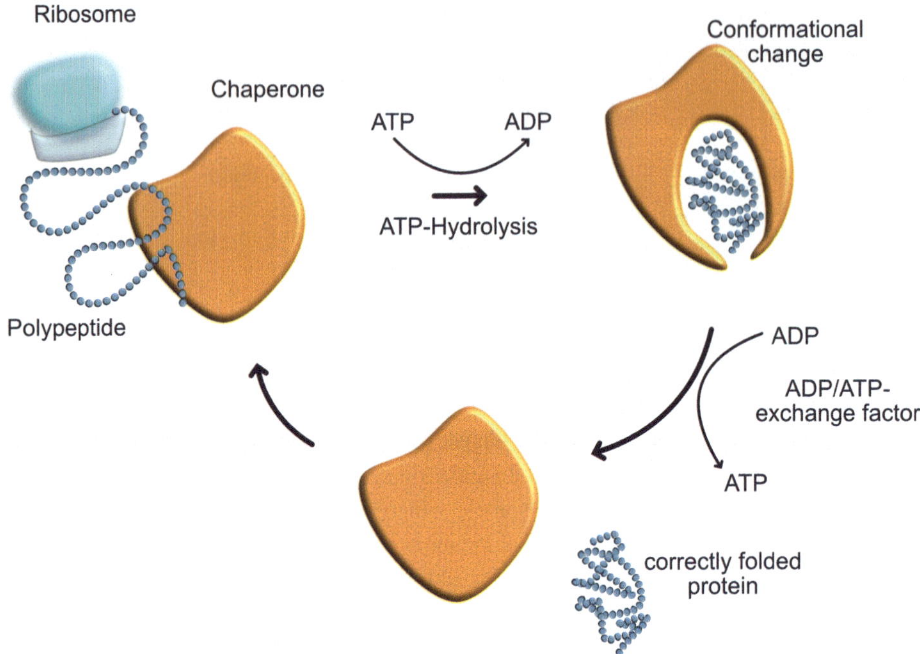

Fig. 2.22 Schematic illustrates function of chaperones, polypeptide chain formed at ribosome is immediately bound by chaperones in the cytoplasm, by hydrolyzing the phosphoanhydrid bond of ATP, energy is released, which induces conformational change in chaperone protein to form a "cave," which restricts space for polypeptide chain and forces it to fold into its 3-dimensional shape, correctly folded protein is released from chaperone, chaperone-ADP is exchanged for ATP, and the next polypeptide chain can be folded

but also by other cellular stress conditions that result in changes of protein homeostasis in cells, including heavy metals, alcohol, hypoxic conditions and others (Ritossa, 1962; Tissieres et al., 1974; Lewis et al., 1975; Evgen'ev, 2021). *H*eat *s*hock *p*roteins (*HSP*) consist of five families, named according to their molecular mass: Hsp100, Hsp90, Hsp70, Hsp60, and "small heat shock proteins." Many family members have dedicated functions in different cell organelles. Chaperones are needed for folding freshly synthesized proteins, sheltering proteins from unwanted interactions and conformational changes, and refolding damaged proteins, e.g., after heat shock. These activities require energy, which is provided by ATP hydrolysis. ATP-bound Hsp70, for instance, binds to nascent amino acid chains as they leave the ribosome during translation. By hydrolyzing ATP, they gain energy, which infers a drastic change in conformation and allows the chaperone to embrace the unfolded protein chain, thereby restricting the space it can occupy. This forces the protein client into its correct (or energetically favored) shape. When another ATP binds, the embrace is lifted and the folded protein is released, allowing another cycle of protein folding with the next amino acid chain (Fig. 2.22). Cofactors are needed for ATP hydrolysis and also for the exchange of ADP for a new molecule of ATP.

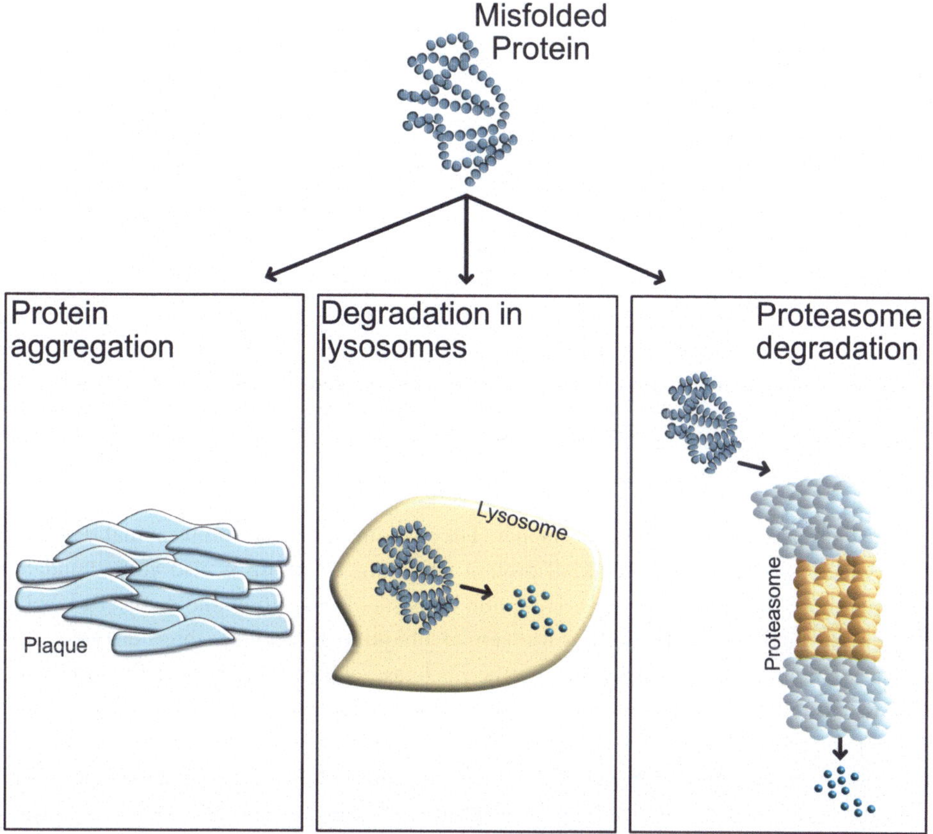

Fig. 2.23 Fates of misfolded proteins that cannot be refolded are illustrated. They include aggregation, degradation in lysosomes, or degradation by the proteasome

During the life of a cell, misfolding of proteins is always possible. Sometimes, proteins can not be refolded. In this case, they await three different fates. Fate 1 is aggregation, which can lead to insoluble protein accumulations of rather large sizes that at some point may even kill the cell. Sometimes, such aggregations of proteins cause disease, for instance, plaque formation in Alzheimer disease. Fate 2 is direction into lysosomes followed by rapid degradation. Finally, fate 3 leads into the proteasome, where misfolded proteins are degraded (Fig. 2.23).

The proteasome is a large protein complex representing a tube. It consists of a 20 S-core and two 19 S-caps covering the tube-like lids (S refers to Svedberg centrifugation constant as explained in Chap. 1). Threonine proteases are present in the core and their active sites are directed toward the inner space of the core tube. Like a shredder, they degrade proteins arriving inside the tube (Budenholzer et al., 2017). Proteasomes are very abundant with an estimated 30,000 proteasomes per cell. Proteins are directed into the

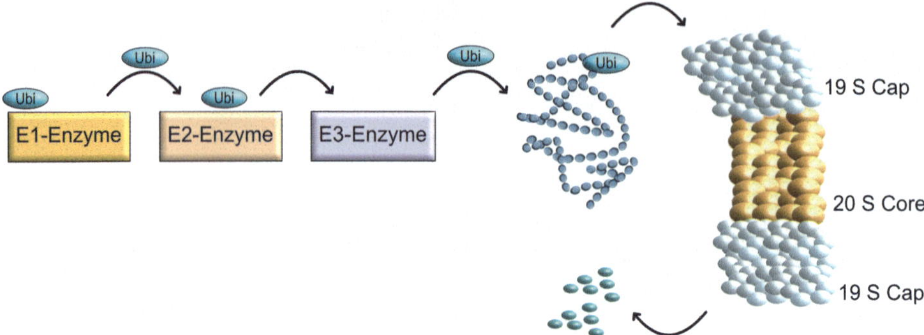

Fig. 2.24 Protein degradation by proteasome: E1-enzyme prepares ubiquitin for transfer onto target protein, E2-enzyme bridges ubiquitin and E3-enzyme by binding to both, E3-enzyme binds substrate protein and catalyzes ubiquitin ligation onto substrate, and ubiquitinylated protein is transferred into proteasome and degraded

proteasome in a strictly regulated manner. They obtain a label for proteasomal degradation, which is provided by the small protein ubiquitin. Ubiquitin is also very abundant, "ubiquitously" distributed in cells. Ubiquitin consists of only 76 amino acids. To label a protein for degradation by adding a chain of ubiquitin molecules, three enzymes are needed. The final step is carried out by the E3 ligase, which directly transfers ubiquitin onto the protein to be degraded, the so-called substrate. E3 cannot just "take" the ubiquitin protein and hand it over to the substrate. Two additional enzymes are required to transfer ubiquitin from its unbound form to E3. E1 activates ubiquitin by forming a thioester bond with the C-terminal glycine carbonic acid group of ubiquitin. Thioester bonds are especially suitable for transferring one amino acid carboxyl to another cysteine; in this case, it is the cysteine of the E2 enzyme, which forms a new thioester with the glycine carbonic acid end of ubiquitin. E3 bridges ubiquitin-conjugated E2 with the substrate and catalyses the formation of a peptide bond between the ubiquitin C-terminal –COOH and the ε-amino group of lysine residues in the substrate (Fig. 2.24).

Some E3 ligases recognize and bind to hydrophobic patches on wrongly folded proteins. In the hydrophilic environment of the cytoplasm of the cell, hydrophobic amino acids are usually hidden inside the protein fold. However, when the proteins are unfolded, those regions get exposed at the surface, making the misfolded protein a substrate for ubiquitination by E3. Usually, a whole chain of ubiquitin residues is attached to the substrate protein to label it for transport into the proteasome, where it is chopped up. Animals have only two E1 enzymes, but ca. 30 E2 enzymes (11 in yeast). The most diverse group of enzymes in the ubiquitination pathway are E3 ligases with about 600–700 members (Steger et al., 2022). Many E3 ligases work in a very specific manner by interacting with only a few or even only one substrate protein. Eradication of unfolded proteins is only one function of E3 ligases. These enzymes are also involved in regulating specific cellular pathways by degrading pathway components when required, for instance, during the cell cycle or as reaction to environmental damage. Some proteins have specific degradation

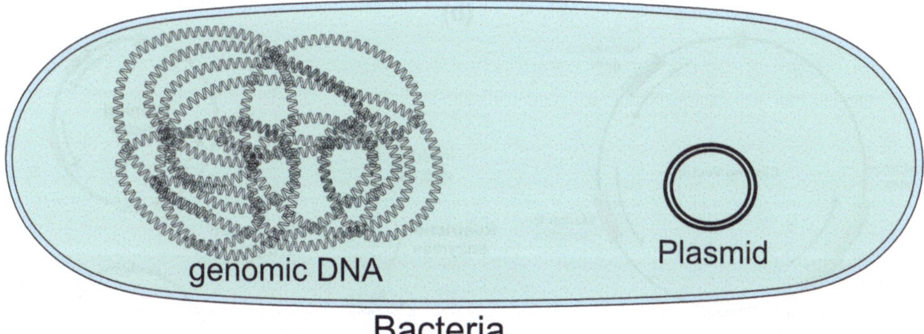

Fig. 2.25 Schematic representation of bacterium with genomic DNA and plasmid DNA

signals, so-called degrons, which constitute recognition signals for E3 ligases (see Chap. 4, mitosis). The proteasome was discovered by several researchers between 1977 and 1980, and Aaron Ciechanover, Avram Hershko (Goldknopf & Busch, 1977; Ciechanover et al., 1980) and Irwin Rose (Hershko et al., 1980) received the Nobel Prize for their discoveries in 2004.

2.2.6 Genetic Engineering

A major contribution to our current understanding of cellular functions has been provided by methods of genetic engineering that were developed on the basis of the knowledge about heredity of DNA, including DNA replication mechanisms and the flow of genetic information from DNA via mRNA to the ribosome for translation of mRNA into proteins. The basic tools for genetic engineering are plasmids. Plasmids are small rings of double-stranded DNA, which are found in bacteria in addition to the chromosomal genomic DNA (Fig. 2.25). With their own origin of replication, they are replicated autonomously for each cell division. They carry genetic information for certain bacterial characteristics. One example is the fertility factor (*F*), which controls bacterial conjugation, a parasexual bacterial behavior. It can be exchanged between different individuals of a bacterial colony.

Original bacterial plasmids were reconstructed by scientists to be usable as vectors to move certain gene sequences between different organisms and even species. A typical plasmid vector contains such a gene sequence (also called insert) and a promoter sequence, which is specific for the organism where this gene is supposed to be expressed. The gene sequence is flanked by specific short palindromic DNAse recognition sites, so-called restriction sites. By using the appropriate sequence-specific DNAse (the restriction enzyme), the plasmid can be cut at specific positions (Fig. 2.26a). There is a huge toolbox of restriction enzymes, each of which has a unique DNA cleavage site. After cleavage of the DNA sequence on both strands, in most cases, overhangs ("sticky ends") are produced, which allow either re-ligation of the DNA at this position or insertion of a piece of double-

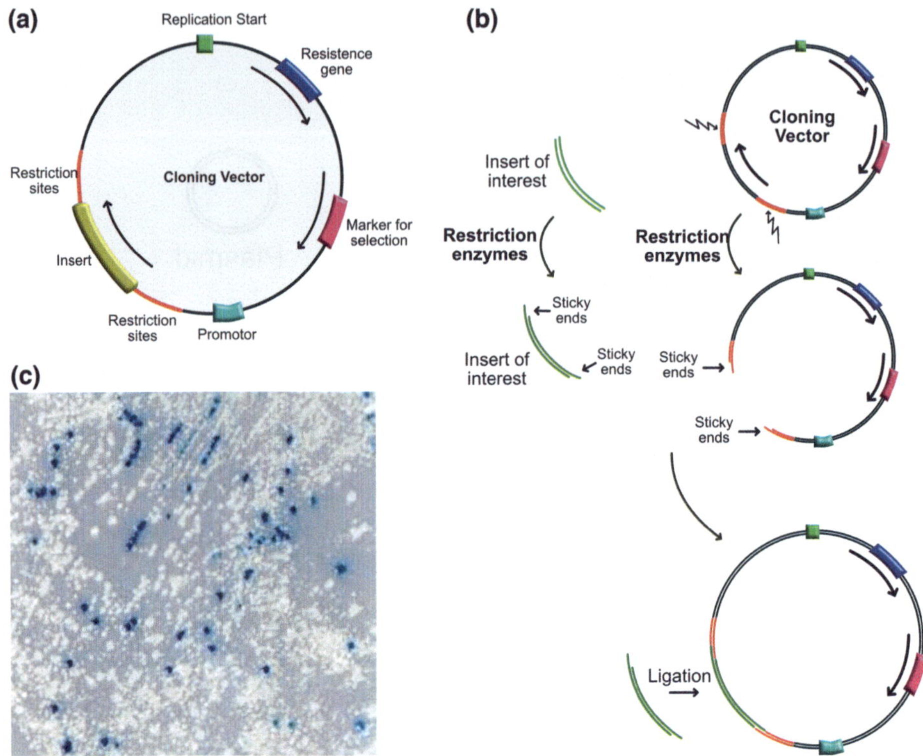

Fig. 2.26 (**a**) Schematic representation of cloning vector with replication start site, DNA sequence coding for an antibiotic resistance, DNA sequence encoding selection marker, promoter sequence-specific for certain organisms, restriction sites, and an DNA insert encoding a desired protein sequence; (**b**) cloning process includes restriction enzyme cutting of the vector at specific DNA sequences (restriction sites) leaving DNA overhangs at both strands, so-called sticky ends, insert DNA of interest is produced by PCR amplification and also digested with restriction enzymes leaving sticky ends matching those of the vector, insert DNA and cloning vector are fused together by the enzyme DNA ligase; (**c**) example for blue–white selection of bacteria carrying the desired plasmid: When plasmid is present and contains selection marker gene in the correct reading frame, the enzyme β-galactose is produced and generates a blue-colored product when incubated with the substrate X-Gal. This indicates either insert incorporation or nonincorporation, depending on the position of the marker on the plasmid. Bacterial colonies containing the plasmid with the desired insert can be easily identified

stranded DNA that has been cleaved by the same enzyme (Fig. 2.26b). Moreover, researchers inserted DNA sequences encoding selectable markers into vectors, e.g., certain enzymes the presence of which can be detected by adding a substrate. An example for such a marker enzyme is β-galactosidase (also called lactase or β-gal for short), which catalyzes the hydrolysis of β-galactosides (e.g., lactose) into monosaccharides (see later). When an artificial substrate (*chromogenic substrate 5-bromo-4-chloro-3-indolyl-β-d-galactopyranoside, X-gal*) is added to the enzyme, hydrolysis of the glycosidic bond results in one molecule of galactose and an intermediate, which turns into an intense blue product by dimerization and oxidation. Thus, bacterial colonies expressing this enzymatic marker turn blue upon addition of this substrate, allowing a "blue–white" selection of bacteria carrying this plasmid from such that do not (Fig. 2.26c). Furthermore, plasmid vectors contain antibiotic resistance genes for selection. Bacteria carrying a specific resistance gene, e.g., an ampicillin resistance, can grow on plates with ampicillin, and bacteria without the plasmid cannot.

Insert sequences can be amplified from the DNA of the tissue of interest by PCR (Fig. 2.10). By adding the restriction site sequences to the primers, the final PCR product will terminate with this sequence. The restriction enzyme is then used to digest the PCR product leaving it with the "sticky ends" that allow insertion into the vector cleaved by the same enzyme (Fig. 2.26b). In many cases, genes of interest are amplified from copyDNA obtained from mRNA. mRNA is devoid of intron sequences and thus allows easy amplification of an "open reading frame."

Vectors are amplified in bacteria, usually special laboratory strains of the intestinal bacteria of the species *E. coli*. For this, the bacteria have to take up the plasmid first. Some bacteria have a natural competence for DNA uptake, which they, for instance, use to exchange plasmids during natural conjugation. However, bacteria of *E. coli* take up DNA also when a short heat or electrical treatment is applied. However, they have to be pretreated with salt solutions to obtain the "competence" for this process. Nonviral transfer of DNA into competent bacteria, fungi, or plants is called "transformation," transfer of cloning vectors into animal cells is called "transfection." For cloning and amplification of plasmids, only bacteria can be used. The plasmid DNA can then be isolated from bacteria (separate from their genomic DNA) and cells of all species can be either "transformed" with it or it can be "transfected" (according to the definitions of the process of DNA transfer). Successfully transfected cells will eventually express the transgene (the artificially inserted gene encoded on the plasmid).

An example of a cloning vector is pcDNA3 (Fig. 2.27). It contains 5446 base pairs, the f1 origin of replication, which is derived from the bacterial fertility factor plasmid. It also has a promoter sequence for expression of genes in human cells. This sequence is taken from the human cytomegalovirus (CMV), a virus of the Herpes virus family. Viruses have often evolved promoter sequences that allow very efficient transcription of their own genes in human cells. The vector additionally contains an antibiotic-resistance gene for selecting transformed bacteria.

Plasmid construction and transformation or transfection into target cells is the basis for ectopic gene expression. Genes can be ectopically (outside of their typical endogenous context) expressed in bacteria, for instance. Bacteria then produce the protein encoded by these transgenes. These protein products are often purified from large bacterial colonies and used for biochemical and structural investigations. But it is also possible to ectopically express genes in plants or animals and study their function in a specific organic context.

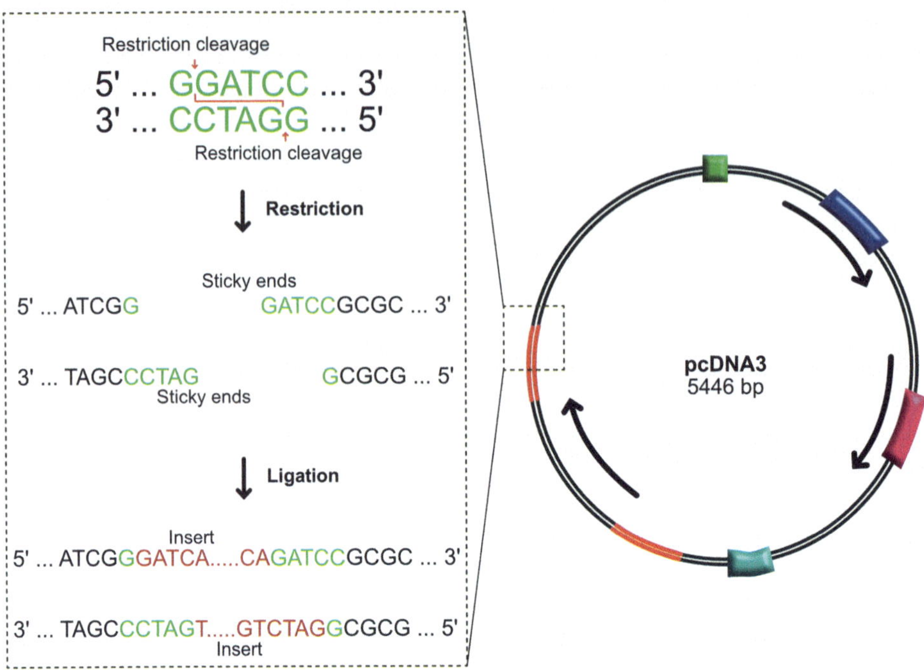

Fig. 2.27 Example of DNA sequence to be cut by restriction enzyme leaving "sticky ends" (green). After restriction, these ends match with sticky ends of PCR-product of insert DNA with added restriction site sequences, after cleavage with the same enzyme (red). This allows effective annealing and ligation of insert and cloning vector

When investigating the function of human genes, often, cell cultures are used. For establishing a cell culture, animal tissue is homogenized into cells by mechanical or enzymatic processing. This results in cell suspensions, which can be grown in Petri dishes. Cells that naturally adhere to the extracellular matrix, such as epithelial cells, will also adhere to plastic dishes, divide, and grow into dense lawns. This will result in a so-called primary culture of adherent cells, which can repeatedly be suspended from the lawn and diluted to be seeded again in new dishes through the process of "passaging." Passaging of a primary culture derived from healthy tissue can only be carried out for a certain number of times, then the cells die. Many experiments can be carried out with such primary cultures for a limited time. Some cells, such as fibroblasts, can be immortalized. After a certain number of passages, most of the fibroblasts in such a culture experience a so-called crisis and die. However, some remain. These can then be propagated indefinitely. Another way to obtain permanently dividing cells is transfection with genes encoding tumor proteins, so-called oncogenes. These make cells independent from cell cycle control mechanisms and thus inevitable. Primary cultures taken from tumor tissue have the same properties and are usually able to grow for indefinite numbers of passages. Many immortal and tumor cell lines are stored at tissue culture collections and distributed among researchers worldwide (ATCC).

Transfection of animal cells can be carried out chemically or physically. The latter involves three commonly used techniques, including (1) injecting DNA directly into the

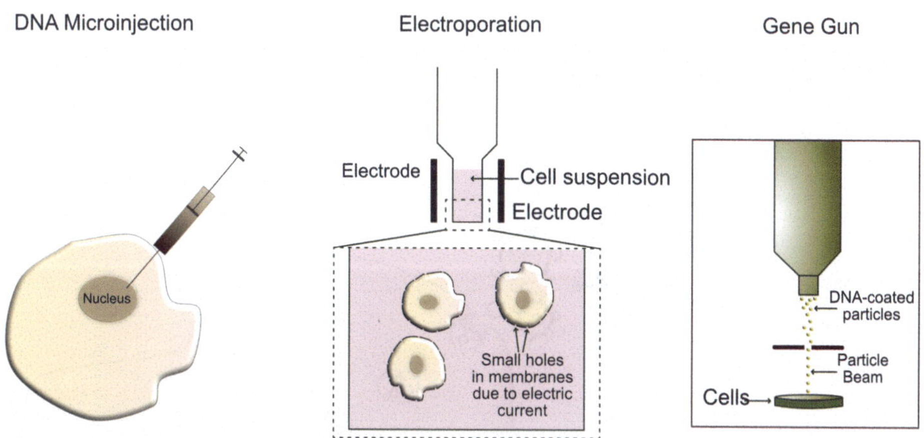

Fig. 2.28 Physical incorporation of ectopic DNA into cells including microinjection; Electroporation and ballistic transfection by gene gun

nucleus of the cells, which is very tedious, (2) when cells are exposed to a brief electrical shock, cell membranes obtain small holes for a short moment where DNA can enter, and (3) also, biolistic methods are available, for instance, a "gene gun," where DNA-covered gold particles are shot at the cells or tissues with strong pressure (Fig. 2.28).

For chemical transfection, three major technologies are used (Fig. 2.29). (1) The DNA is mixed with calcium phosphate, which leads to the formation of precipitates that are ingested by the cell by endocytosis. (2) Cationic polymers such as polyethylenimine (PEI) are used to wrap up the DNA. The PEI–DNA complex is positively charged and when it engages with the negatively charged parts of the cell membrane it is taken up by endocytosis. DNA will then escape from the endosome by a mechanism probably involving osmotic rupture of the endosome. When the cells divide, the DNA will be able to enter the nucleus where it can be transcribed into mRNA, the first step of expressing the ectopic gene. (3) DNA is mixed with specific lipids and is thereby enclosed into liposomes, which fuse with the cell membrane and deliver the DNA into the cell.

This technique was recently refined by developing lipid nanoparticles (LNP) and is used for gene transfer into humans. The history of introducing mRNA into laboratory animals or into humans began in 1989 when Robert Melone discovered that lipid droplets allowed RNA transfection. Around the year 2000, two companies were founded, which wanted to use this technology for vaccination, Curavec and Biontech, both in Germany. It was soon discovered that RNA transfections into mammals caused massive infections, which then led the researchers to exchanging pseudouridine for uridine in such RNAs. This was picked up by the US company, Moderna. Lipid nanoparticles (LNPs) were developed from 2012 on. Clinical trials based on RNA transfection by LNPs had never really been successful with only one clinical trial reaching phase 2. Despite this situation, this technique was and is to this date used on billions of people to vaccinate against SARSCoV-2 (Dolgin, 2021) (see Box 2.2). Unfortunately, recent analyses revealed a significant number

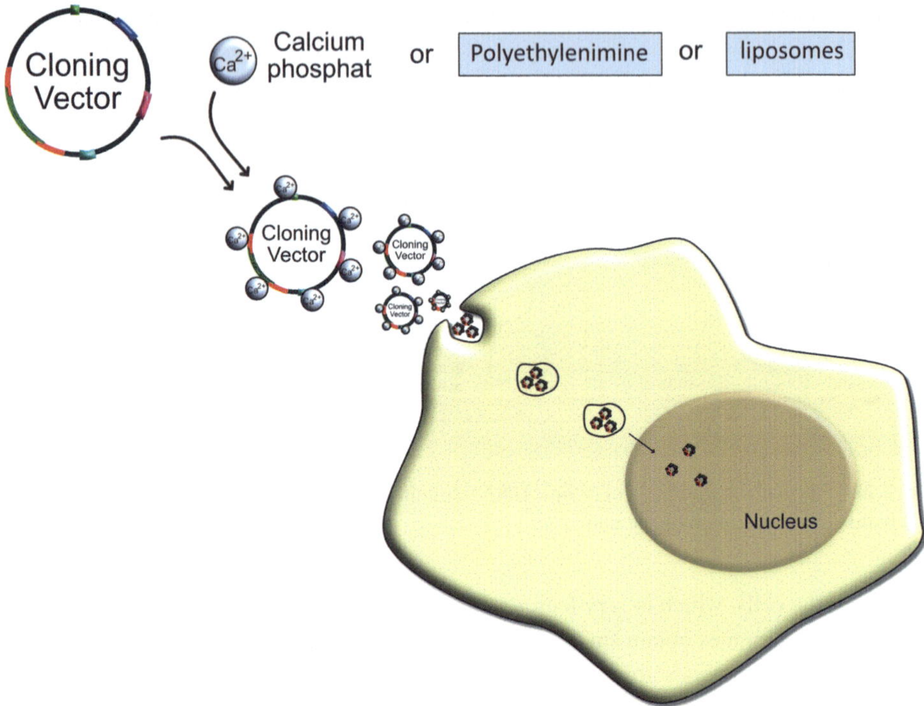

Fig. 2.29 Chemical incorporation of ectopic DNA into cells, using calciumphosphate–DNA complexes, using charged polyethylenamine–DNA particles or using lipid droplets

Box 2.2 DNA and RNA Vaccines

Most of the recently developed vaccines against the SARSCoV 2 virus essentially use a gene transfection strategy to make human cells produce the spike protein after ectopic expression of the gene. The so-called mRNA vaccines (Biontech and Moderna) transfect RNA into the human body. As transfection agent, lipid nanoparticles are used. They contain a lipid bilayer consisting of a polyethylenglycol lipid, a charged ionizable lipid, a neutral ionizable lipid, cholesterol and a normal lipid. Under pressure, these components are mixed with the mRNA, which will then be enveloped by the lipid bilayer. The lipid nanoparticles with RNA can basically enter every cell of the human body. The DNA vaccines (Johnson and Johnson and Astra Zeneca) use the envelope of the adenovirus as vehicle to transport DNA encoding spike protein into the human cells. Adenovirus vaccines should enter natural target cells of adenoviruses, such as epithelia cells.

of adverse events of special interest (AESI) with mRNA vaccines, including myocarditis, pericarditis, and Guillain-Barré syndrome (Faksova et al., 2024; Hu et al., 2024). Similar safety signals were obtained for adenovirus-based vaccines. Whether they are due to a general reaction of the human body to genetic modification, the widespread distribution of the transfection agents in the human body, or by impurities in the individual charges of the preparations has not yet been clarified.

The gene expressing a green fluorescent protein (GFP) is probably the one most often transfected into cells these days. This protein was first discovered in the medusae *Aequorea Victoria* living in very deep sea waters. When it is excited with blue light (that the medusae produce by chemiluminescence at certain stimuli), the umbrella of the medusae emits pulses of green fluorescent light. GFP is a relatively small protein of 238 amino acids. None of the amino acids emits fluorescence; however, when the protein is folded into its typical barrel structure, in the middle of the barrel, three amino acids come into close contact and build a fluorophore using oxygen. Often, the gene sequence encoding GFP is joined with the sequence of another protein-encoding sequence. As a result, a so called "fusion protein" or "GFP-tagged" protein will be expressed in the transfected cell and can be observed using fluorescence microscopy (Fig. 2.30a). Figure 2.30b shows a *Hydra* polyp expressing GFP in some of its cells and a *Hydra* cell expressing a GFP protein-tagged protein of the outer mitochondrial membrane (Fig. 2.30b, c).

An important prerequisite for all techniques described here is the possibility for sequencing DNA. At present the genomes and transcriptomes of many species have been sequenced. The best known method for DNA-sequencing is the so-called Sanger-sequencing. Sanger was the first to determine the amino acid sequence of insulin in 1955. For this he had to remove one amino acid after the other from the insulin amino acid chain and define its identity by thin layer chromatography. He obtained the Nobel prize for this achievement. However, he also developed a method for DNA-sequencing, which soon took over the whole field and was awarded with another Nobel prize. His "chain termination" method uses compromised radioactively labelled nucleotides to be incorporated into the DNA during DNA-strand synthesis by DNA-polymerase. These nucleotides have their free –OH removed from the 3′-position of the sugar, making them di-deoxyribonucleotides (Fig. 2.31a). Thus, when such a nucleotide is incorporated into the nucleotide chain, the polymerase cannot add another nucleotide and the synthesis is terminated. With the DNA to be sequenced, four reactions have to be performed, each including a primer, for the DNA polymerase to start synthesis, the DNA polymerase, and all four nucleotides, whereby for each reaction, a certain amount of a radioactively or fluorescently labeled di-deoxyribonucleotide is included. Reaction 1 receives labeled di-deoxyadenosine, reaction 2 labeled di-deoxyguanosine, reaction 3 labeled dideoxythymidine, and reaction 4 labeled di-deoxycytosine. DNA synthesis then yields DNA chains of different lengths, which always end with the labeled di-deoxynucleotide in each reaction. With radioactive labeling techniques, the DNA pieces are subsequently separated on a polyacrylamide gel by electrophoresis, whereby the DNA products of each reaction are run on separate lanes. All strands distributed over four lanes in the gel will then differ in their length by one

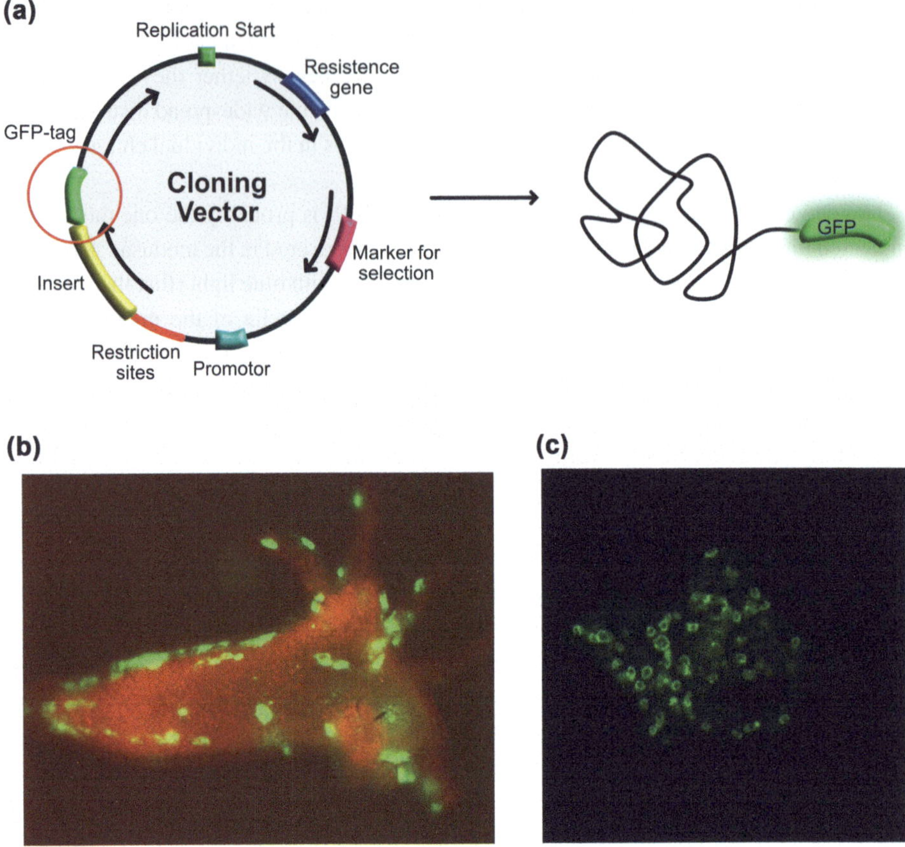

Fig. 2.30 (**a**) Example of cloning vector encoding the protein sequence of the green fluorescing protein (GFP) alone, or fused to the coding sequence of a protein of interest providing a tag; (**b**) fresh water polyp after ballistic transformation with a cloning vector encoding GFP; single epithelial cells show GFP expression; (**c**) human cell transfected with cloning vector encoding protein of the outer mitochondrial membrane fused to GFP

nucleotide. The gel can be "read" by adding the bases at the end of each strand in steps of one nucleotide size difference (Fig. 2.31b). When fluorescence is used, each of the four di-deoxyribonucleotides receives a different "color." The four reactions can then be separated within one lane of a "capillary" gel electrophoresis and red, green, yellow, and blue fluorescence can be red with a detector after laser excitation. With these methods, DNA pieces of ca. 900 bp can be sequenced. Large DNA-chains such as those represented by chromosomes therefore have to be digested into smaller pieces. After sequencing those, sequence overlaps have to be found in order to assemble the correct sequence of the whole

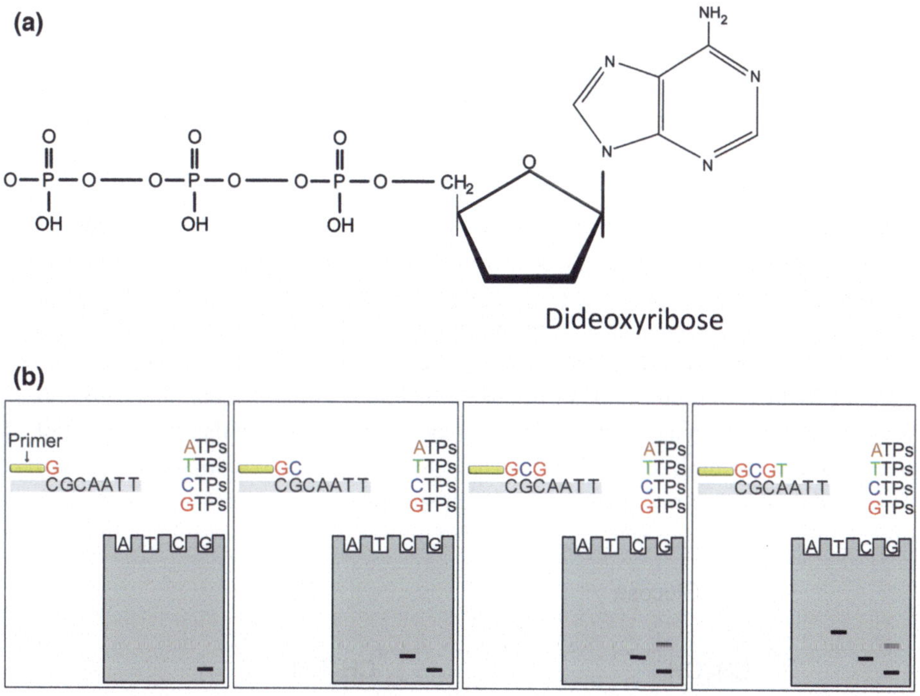

Fig. 2.31 Sanger sequencing; (a) structural formula of di-deoxyribose lacking the 3′ hydroxyl group, it cannot be linked to 5′ phosphate by DNA polymerase; (b) gel electrophoresis of DNA strands that have incorporated radioactively labeled di-deoxyribose stop nucleotides at different lengths (1, 2, 3, or 4), samples treated with A, T, C, or G-stop nucleotides are separated on different lanes

chromosome. This method was used for sequencing the first human genome (Venter et al., 2001).

In the years to come, DNA-sequencing methods experienced a revolution. So called "next-generation sequencing" methods were automated fast and became affordable procedures. On glass plates, or chips, many chains of DNA were now sequenced in parallel. This resulted in a huge increase of available genome sequences. The actual pieces that are sequenced are often very short with a maximum of 200 base pairs, often under 100. For such sequences to be assembled, a template is required. Sequencing therefore is strongly coupled with bioinformatic techniques. It is, for instance, possible to sequence the two billion base pairs of a human person relatively fast by such a method. For assembly of all the short sequences in the output of the reaction, they are simply mapped on the existing human genome, which is the totality of all DNA sequences identified in humans.

2.3 Carbohydrates

The building blocks of polysaccharides are monosaccharides such as glucose or fructose. Monosaccharides are *carbohydrates* representing covalently bound carbon with water $(CH_2O)n$. Monosaccharides are joined by *glycosidic bonds* forming an ether between the hydroxyl group of one and the anomeric carbon of the next by a condensation reaction releasing water (Fig. 2.32). They are very diverse due to the number of possible monosaccharides including glucose, fructose, galactose with 6 carbon atoms each, or ribose with 5. Moreover, each hydroxyl group can in principle form glycosidic bonds with another sugar molecule whereby a large variety of linear and branched polysaccharides can be formed. Finally, carbohydrate building blocks are present in different stereoisomeric versions where the hydroxyl group can be positioned either above or underneath the carbon ring. Two stereoisomers then have equal atomic composition but differ in the special position of certain parts of the molecules. Thereby, each stereoisomer represents itself as a mirror-image of the opposing one, making two such stereoisomers mirror-isomers or enantio-

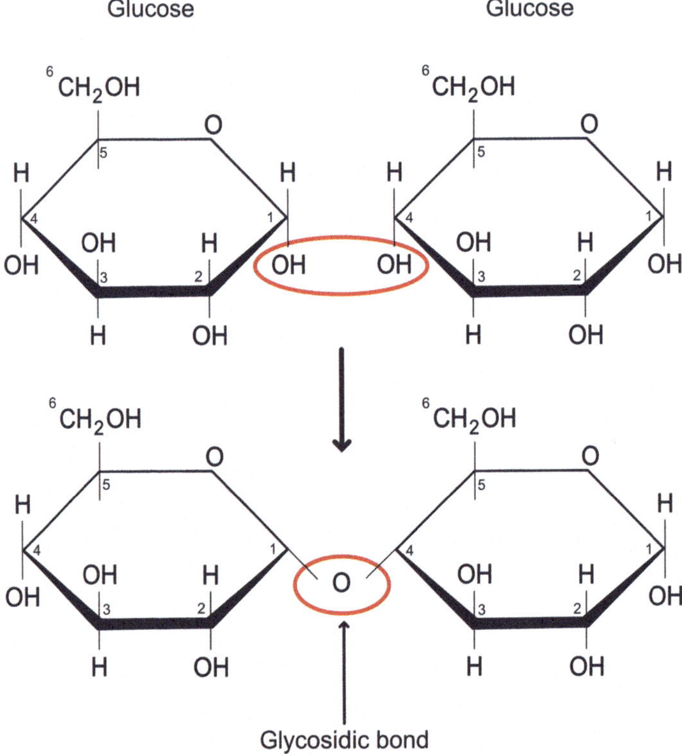

Fig. 2.32 Structural formula of glucose; 2 glucose molecules form a glycosidic bond; 2 hydroxyl groups at position 1 of first glucose and position 4 at second glucose in upper panel, engage in 1,4 glycosidic bond in lower panel as indicated by red circles

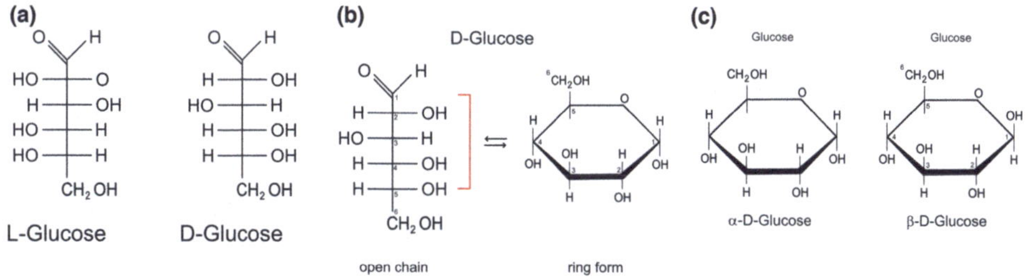

Fig. 2.33 (**a**) Structural formulas of L- and D-glucose; (**b**) open-chain- and ring configurations of D glucose; (**c**) α-and β-D glucose

mers. This is shown in Fig. 2.33a with D- and L-glucose as an example. Moreover, D-glucose occurs in two isoforms again, the 1′ OH-group can either point downward from the carbon ring (α-D-glucose) or upwards (β-D-glucose) (Fig. 2.33c). Accordingly, α- or β-glycosidic bonds can be formed with the 4′ OH-group of another molecule of D-glucose. In its solid form, glucose represents an open chain, whereas in water, it forms a ring (Fig. 2.33b).

Glucose is formed in plants from water and CO_2 during photosynthesis using the energy of the sunlight. It is the major energy source of all living beings and it is stored as a component of large polysaccharides, glycogen in animals, and starch in plants. Moreover, plant cell walls consist of cellulose. Cellulose is a long chain of β-D-glucose (Fig. 2.34a), whereas glycogen is formed by branched chains of α-D-glucose (Fig. 2.34b). Glycogen is stored in granules in liver and muscle cells. These granules have a core of the protein glycogenin. Starch is the major storage form of glucose in plants and it consists of the linear polysaccharide amylose and the branched polysaccharide amylopectin (Fig. 2.34c). The basic units are in both cases maltose disaccharides (Fig. 2.35) and they are connected by 1,4 or for branching by 1,6 α-glycosidic bonds. Here, the Starch grains range in size from 3 to 180 μm and can only be made visible by microscopic analyses. They differ in their shape and size between different plant species. Simple starch grains are shaped like an ellipse, but there are also starch grains that are composed of several subgrains and they can adapt complicated polygonic forms, almost like crystals.

Two sugar molecules form a disaccharide, for instance, maltose, which is composed of two molecules α-D-glucose. Maltose is a degradation product of starch. Lactose is present in milk and is composed of galactose and glucose. Sucrose, our household sugar, is yielded from sugar beet or sugar cane, and composed of glucose and fructose (Fig. 2.35). Interestingly, glucose, fructose, and galactose have the same sum formula $C_{12}H_{22}O_{11}$, but are differently structured.

Polysaccharides are often attached to proteins specifically to hydroxyl groups of serine, threonine, and tyrosine (O-linked polysaccharides) or to amino groups of lysine or arginine residues (N-linked polysaccharides). Attachment of polysaccharides to proteins occurs in the ER and Golgi apparatus by enzymes called glycosyltransferases. In addition, polysaccharides attach to lipids, producing glycolipids. Glycoproteins or glycolipids are present

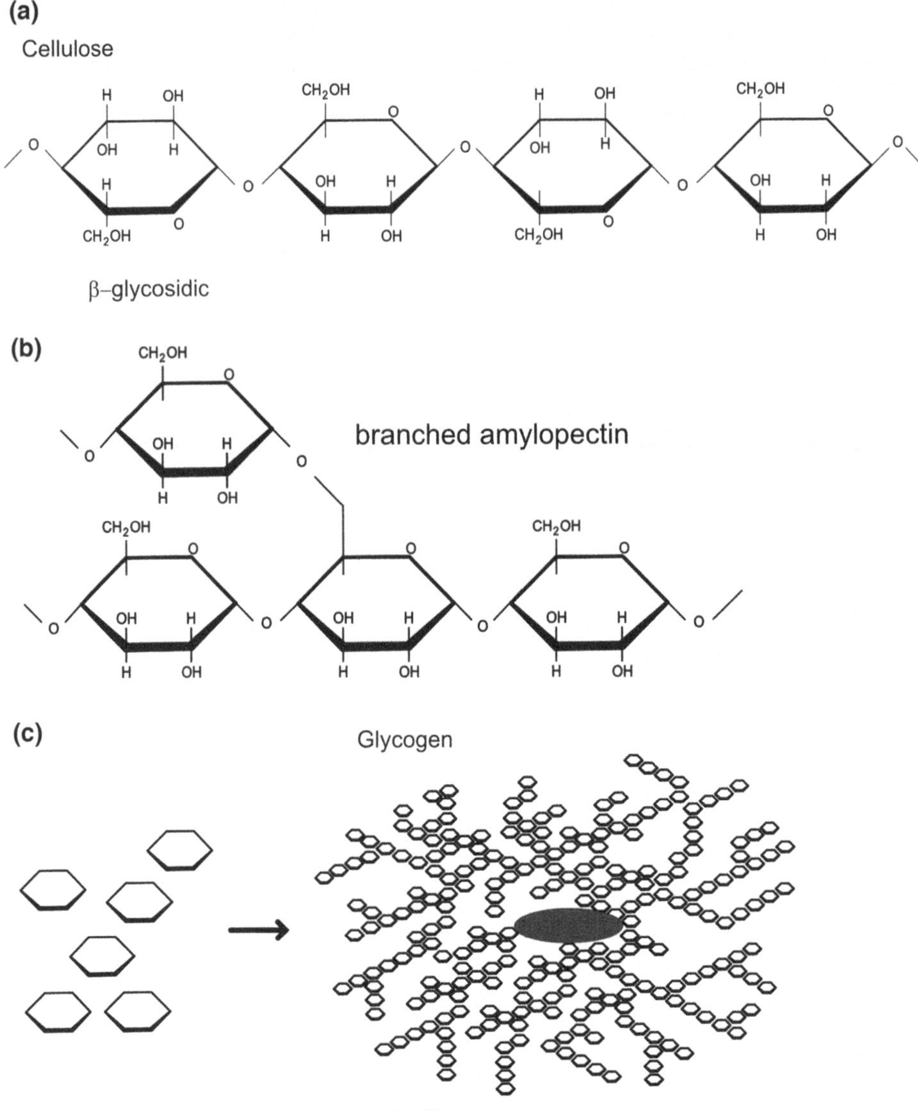

Fig. 2.34 (**a**) 1,4 glycosidic bonds between molecules of D-glucose chains as present in cellulose; (**b**) 1,4 and 1,6 glycosidic bonds of D-glucose molecules in amylopectin lead to branched structure; (**c**) schematic representation of glycogen, where D-glucose forms 1,4, and 1,6 glycosidic bonds, resulting in branched structure; this is arranged around a protein core containing the glycogenin

at the cell surface of all cells and confer a very specific signature to each cell type (Fig. 2.36). They often form very elaborate extracellular structures, a so-called sugar coat, or glycocalyx, which surrounds all cells in unicellular as well as multicellular organisms. Due to the diversity of polysaccharides, the "sugar coats" of cells are also very diverse and

Fig. 2.35 Structural formulas of di-saccharides maltose, saccharose, and lactose

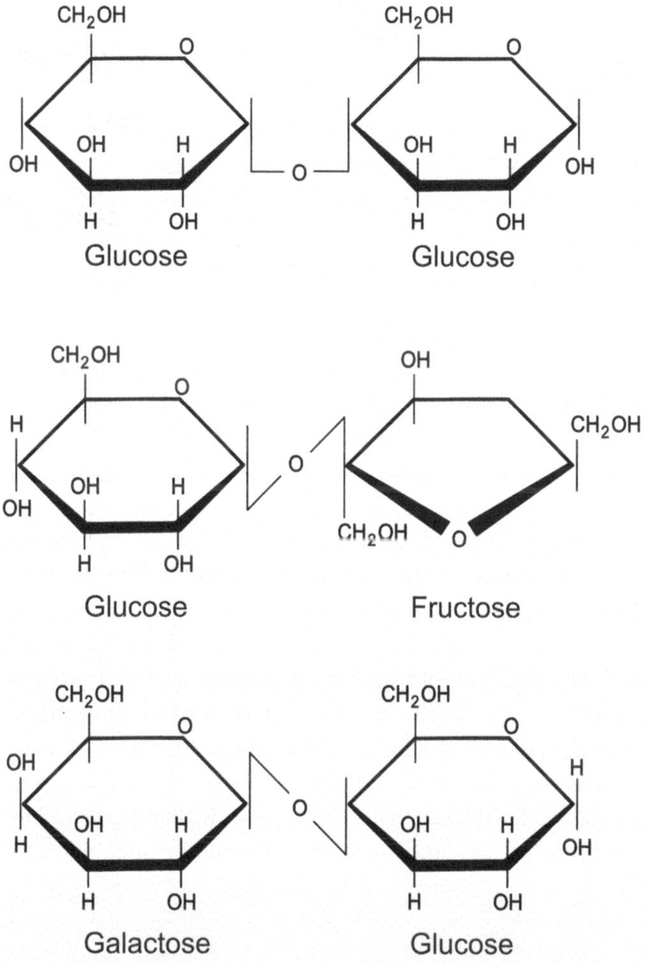

often are extremely cell-specific to the extent that they can be used by an organism to distinguish between its own and foreign cells. This is, for instance, also the basis for our different blood groups in the AB0-system. The antigens that define these blood groups are glycoproteins on the surface of red blood cells. If blood samples of two different blood groups, e.g., from different people are mixed, the erythrocytes clump together or agglutinate because of anti-blood group antigen antibodies that are present in the serum and react with foreign erythrocytes, but not with self. A specific class of proteins is able to interact with carbohydrates, and these are lectins.

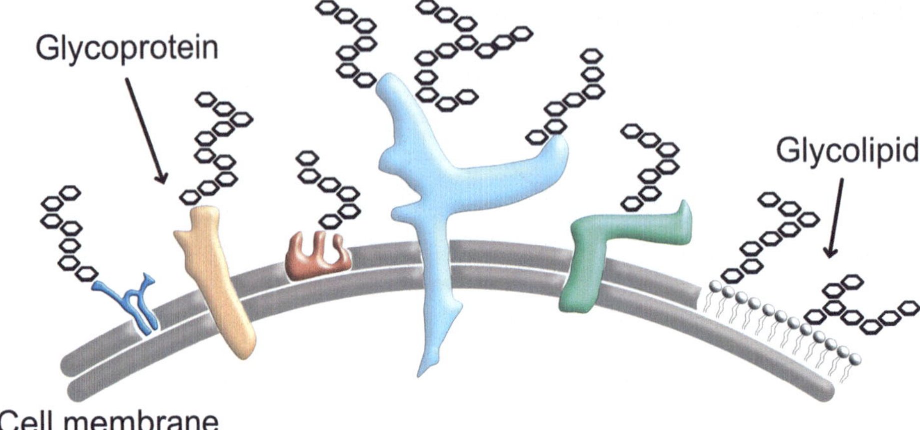

Fig. 2.36 Schematic representation of cell membrane with "sugar coat." Sugar hexagons are illustrated as they are attached to membrane proteins (colored) and lipids of the cell membrane

2.4 Lipids

The building of lipids, the last group of macromolecules discussed here, follows a different logic. Lipids are esters of fatty acids with the trimeric alcohol glycerine. They are also called triglycerides. Their content of fatty acids makes them insoluble in water, and they are hydrophobic or lipophilic. Fatty acids are long carbon chains with a COOH-group at their end that can be either saturated when each carbon is bound to two hydrogens, or unsaturated, when hydrogens are lacking and double bonds are formed between two carbon atoms. An important subfamily of lipids are phospholipids. Here, glycerol forms an ester with only two fatty acids and the third hydroxyl forms an ester with phosphate. This makes phospholipids apolar and polar at the same time and they form the lipid bilayer of our membranes (Chap. 3).

Test Yourself

1. Name the four classes of macromolecules that are needed for cell function.
2. How are they composed? Which chemical bond keeps the monomers together?
3. Explain the flow of genetic information in a cell from DNA via RNA to protein synthesis.
4. What are the primary, secondary, tertiary, and quaternary structures of proteins?
5. How are nucleic acids introduced into cells?
6. How is DNA replicated?
7. What does "genetic engineering" mean?
8. What are mRNA vaccines?
9. What are carbohydrates?
10. What is the chemical composition of lipids?

References

Allfrey, V. G., Faulkner, R., & Mirsky, A. E. (1964). Acetylation and methylation of histones and their possible role in the regulation of RNA synthesis. *Proceedings of the National Academy of Sciences of the United States of America, 51*(5), 786–794.

ATCC. Available from https://www.atcc.org

Baltimore, D. (1970). RNA-dependent DNA polymerase in virions of RNA tumour viruses. *Nature, 226*(5252), 1209–1211.

Bannister, A. J., & Kouzarides, T. (2011). Regulation of chromatin by histone modifications. *Cell Research, 21*(3), 381–395.

Böttger, A., Vothknecht, U., Bolle, C., & Wolf, A. (2018). *Lessons on Caffeine, Cannabis & Co: Plant derived drugs and their interaction with human receptors*. Springer.

Brenner, S. (1999). Theoretical biology in the third millennium. *Philosophical Transactions of the Royal Society of London. Series B, Biological Sciences, 354*(1392), 1963–1965.

Brock, T. D., & Freeze, H. (1969). Thermus aquaticus gen. n. and sp. n., a nonsporulating extreme thermophile. *Journal of Bacteriology, 98*(1), 289–297.

Budenholzer, L., Cheng, C. L., Li, Y., & Hochstrasser, M. (2017). Proteasome structure and assembly. *Journal of Molecular Biology, 429*(22), 3500–3524.

Ciechanover, A., Elias, S., Heller, H., Ferber, S., & Hershko, A. (1980). Characterization of the heat-stable polypeptide of the ATP-dependent proteolytic system from reticulocytes. *The Journal of Biological Chemistry, 255*(16), 7525–7528.

Corman, V., Bleicker, T., Brünink, S., & Drosten, C. (2020a). *Diagnostic detection of Wuhan coronavirus 2019 by real-time RTPCR*. Available from https://www.who.int/docs/default-source/coronaviruse/wuhan-virus-assay-v1991527e5122341d99287a1b17c111902.pdf

Corman, V. M., Landt, O., Kaiser, M., Molenkamp, R., Meijer, A., Chu, D. K., Bleicker, T., Brunink, S., Schneider, J., Schmidt, M. L., Mulders, D. G., Haagmans, B. L., van der Veer, B., van den Brink, S., Wijsman, L., Goderski, G., Romette, J. L., Ellis, J., Zambon, M., Peiris, M., Goossens, H., Reusken, C., Koopmans, M. P., & Drosten, C. (2020b). Detection of 2019 novel coronavirus (2019-nCoV) by real-time RT-PCR. *Euro Surveillance, 25*(3), 2000045.

Cremer, T., Kreth, G., Koester, H., Fink, R. H., Heintzmann, R., Cremer, M., Solovei, I., Zink, D., & Cremer, C. (2000). Chromosome territories, interchromatin domain compartment, and nuclear matrix: An integrated view of the functional nuclear architecture. *Critical Reviews in Eukaryotic Gene Expression, 10*(2), 179–212.

Dolgin, E. (2021). The tangled history of mRNA vaccines. *Nature, 597*(7876), 318–324.

Evgen'ev, M. B. (2021). Heat shock proteins: A history of study in Russia. *Cell Stress & Chaperones, 26*(4), 617–627.

Faksova, K., Walsh, D., Jiang, Y., Griffin, J., Phillips, A., Gentile, A., Kwong, J. C., Macartney, K., Naus, M., Grange, Z., Escolano, S., Sepulveda, G., Shetty, A., Pillsbury, A., Sullivan, C., Naveed, Z., Janjua, N. Z., Giglio, N., Perala, J., Nasreen, S., Gidding, H., Hovi, P., Vo, T., Cui, F., Deng, L., Cullen, L., Artama, M., Lu, H., Clothier, H. J., Batty, K., Paynter, J., Petousis-Harris, H., Buttery, J., Black, S., & Hviid, A. (2024). COVID-19 vaccines and adverse events of special interest: A multinational Global Vaccine Data Network (GVDN) cohort study of 99 million vaccinated individuals. *Vaccine, 42*(9), 2200–2211.

Goldknopf, I. L., & Busch, H. (1977). Isopeptide linkage between nonhistone and histone 2A polypeptides of chromosomal conjugate-protein A24. *Proceedings of the National Academy of Sciences of the United States of America, 74*(3), 864–868.

Hershko, A., Ciechanover, A., Heller, H., Haas, A. L., & Rose, I. A. (1980). Proposed role of ATP in protein breakdown: Conjugation of protein with multiple chains of the polypeptide of ATP-dependent proteolysis. *Proceedings of the National Academy of Sciences of the United States of America, 77*(4), 1783–1786.

Hu, W. C., Chiu, S. K., Yang, Y. F., & Singh, S. (2024). COVID-19 vaccination reporting and adverse event analysis in Taiwan. *Vaccines (Basel), 12*(6), 591.

Laskey, R. A., Honda, B. M., Mills, A. D., & Finch, J. T. (1978). Nucleosomes are assembled by an acidic protein which binds histones and transfers them to DNA. *Nature, 275*(5679), 416–420.

Lewis, M., Helmsing, P. J., & Ashburner, M. (1975). Parallel changes in puffing activity and patterns of protein synthesis in salivary glands of Drosophila. *Proceedings of the National Academy of Sciences of the United States of America, 72*(9), 3604–3608.

Mullis, K. (2000). *Dancing naked in the mind field.* Vintage Books.

Ritossa. (1962). A new puffing pattern induced by temperature shock and DNP in Drosophila. *Experientia, 18*, 571–573.

Schneider, J., & Shilatifard, A. (2006). Histone demethylation by hydroxylation: Chemistry in action. *ACS Chemical Biology, 1*(2), 75–81.

Song, F., Chen, P., Sun, D., Wang, M., Dong, L., Liang, D., Xu, R. M., Zhu, P., & Li, G. (2014). Cryo-EM study of the chromatin fiber reveals a double helix twisted by tetranucleosomal units. *Science, 344*(6182), 376–380.

Steger, M., Karayel, O., & Demichev, V. (2022). Ubiquitinomics: History, methods, and applications in basic research and drug discovery. *Proteomics, 22*(15–16), e2200074.

Tissieres, A., Mitchell, H. K., & Tracy, U. M. (1974). Protein synthesis in salivary glands of Drosophila melanogaster: Relation to chromosome puffs. *Journal of Molecular Biology, 84*(3), 389–398.

Unoki, M., Masuda, A., Dohmae, N., Arita, K., Yoshimatsu, M., Iwai, Y., Fukui, Y., Ueda, K., Hamamoto, R., Shirakawa, M., Sasaki, H., & Nakamura, Y. (2013). Lysyl 5-hydroxylation, a novel histone modification, by Jumonji domain containing 6 (JMJD6). *The Journal of Biological Chemistry, 288*(9), 6053–6062.

Venter, J. C., Adams, M. D., Myers, E. W., Li, P. W., Mural, R. J., Sutton, G. G., Smith, H. O., Yandell, M., Evans, C. A., Holt, R. A., Gocayne, J. D., Amanatides, P., Ballew, R. M., Huson, D. H., Wortman, J. R., Zhang, Q., Kodira, C. D., Zheng, X. H., Chen, L., Skupski, M., Subramanian, G., Thomas, P. D., Zhang, J., Gabor Miklos, G. L., Nelson, C., Broder, S., Clark, A. G., Nadeau, J., McKusick, V. A., Zinder, N., Levine, A. J., Roberts, R. J., Simon, M., Slayman, C., Hunkapiller, M., Bolanos, R., Delcher, A., Dew, I., Fasulo, D., Flanigan, M., Florea, L., Halpern, A., Hannenhalli, S., Kravitz, S., Levy, S., Mobarry, C., Reinert, K., Remington, K., Abu-Threideh, J., Beasley, E., Biddick, K., Bonazzi, V., Brandon, R., Cargill, M., Chandramouliswaran, I., Charlab, R., Chaturvedi, K., Deng, Z., Di Francesco, V., Dunn, P., Eilbeck, K., Evangelista, C., Gabrielian, A. E., Gan, W., Ge, W., Gong, F., Gu, Z., Guan, P., Heiman, T. J., Higgins, M. E., Ji, R. R., Ke, Z., Ketchum, K. A., Lai, Z., Lei, Y., Li, Z., Li, J., Liang, Y., Lin, X., Lu, F., Merkulov, G. V., Milshina, N., Moore, H. M., Naik, A. K., Narayan, V. A., Neelam, B., Nusskern, D., Rusch, D. B., Salzberg, S., Shao, W., Shue, B., Sun, J., Wang, Z., Wang, A., Wang, X., Wang, J., Wei, M., Wides, R., Xiao, C., Yan, C., Yao, A., Ye, J., Zhan, M., Zhang, W., Zhang, H., Zhao, Q., Zheng, L., Zhong, F., Zhong, W., Zhu, S., Zhao, S., Gilbert, D., Baumhueter, S., Spier, G., Carter, C., Cravchik, A., Woodage, T., Ali, F., An, H., Awe, A., Baldwin, D., Baden, H., Barnstead, M., Barrow, I., Beeson, K., Busam, D., Carver, A., Center, A., Cheng, M. L., Curry, L., Danaher, S., Davenport, L., Desilets, R., Dietz, S., Dodson, K., Doup, L., Ferriera, S., Garg, N., Gluecksmann, A., Hart, B., Haynes, J., Haynes, C., Heiner, C., Hladun, S., Hostin, D., Houck, J., Howland, T., Ibegwam, C., Johnson, J., Kalush, F., Kline, L., Koduru, S., Love, A., Mann, F., May, D., McCawley, S., McIntosh, T., McMullen, I., Moy, M., Moy, L., Murphy, B., Nelson, K., Pfannkoch, C., Pratts, E., Puri, V., Qureshi, H., Reardon, M., Rodriguez, R., Rogers, Y. H., Romblad, D., Ruhfel, B., Scott, R., Sitter, C., Smallwood, M., Stewart, E., Strong, R., Suh, E., Thomas, R., Tint, N. N., Tse, S.,

Vech, C., Wang, G., Wetter, J., Williams, S., Williams, M., Windsor, S., Winn-Deen, E., Wolfe, K., Zaveri, J., Zaveri, K., Abril, J. F., Guigo, R., Campbell, M. J., Sjolander, K. V., Karlak, B., Kejariwal, A., Mi, H., Lazareva, B., Hatton, T., Narechania, A., Diemer, K., Muruganujan, A., Guo, N., Sato, S., Bafna, V., Istrail, S., Lippert, R., Schwartz, R., Walenz, B., Yooseph, S., Allen, D., Basu, A., Baxendale, J., Blick, L., Caminha, M., Carnes-Stine, J., Caulk, P., Chiang, Y. H., Coyne, M., Dahlke, C., Deslattes Mays, A., Dombroski, M., Donnelly, M., Ely, D., Esparham, S., Fosler, C., Gire, H., Glanowski, S., Glasser, K., Glodek, A., Gorokhov, M., Graham, K., Gropman, B., Harris, M., Heil, J., Henderson, S., Hoover, J., Jennings, D., Jordan, C., Jordan, J., Kasha, J., Kagan, L., Kraft, C., Levitsky, A., Lewis, M., Liu, X., Lopez, J., Ma, D., Majoros, W., McDaniel, J., Murphy, S., Newman, M., Nguyen, T., Nguyen, N., Nodell, M., Pan, S., Peck, J., Peterson, M., Rowe, W., Sanders, R., Scott, J., Simpson, M., Smith, T., Sprague, A., Stockwell, T., Turner, R., Venter, E., Wang, M., Wen, M., Wu, D., Wu, M., Xia, A., Zandieh, A., & Zhu, X. (2001). The sequence of the human genome. *Science, 291*(5507), 1304–1351.

Watson, J. D., & Crick, F. H. (1953). Molecular structure of nucleic acids; a structure for deoxyribose nucleic acid. *Nature, 171*(4356), 737–738.

The Biomembrane: Between the Cell and its Environment

3

What You Will Learn in This Chapter
All cells are surrounded by biological membranes. These are formed by phospholipids, which arrange in a bilayer where the fatty acids assemble together by hydrophobic interactions in the middle, and the charged head groups point to the aqueous media inside and outside of the cell. Proteins are inserted into this lipid bilayer. By forming channels, receptors, transporters, or other functional units, proteins contribute to diverse membrane functions, which are vital for the cell. The membrane constitutes a diffusion barrier for all hydrophilic molecules and for ions. In this chapter, it will be discussed which molecules can pass the membrane and which cannot. Due to asymmetric distribution of charged molecules at both sides of the membrane and the insulator properties of the membrane lipids, an electrical membrane potential is established in all cells. In excitable cells such as nerve and muscle cells, this potential is used for electrical signal conduction.

Furthermore, we show in this chapter how proteins are co-translationally inserted into ER membranes and how they are sorted to different membrane compartments. The ER, Golgi-apparatus, lysosomes, endosomes and secretory vesicles constitute the secretory compartment of the cell. Components of this compartment travel back and forward within this compartment. All such traveling occurs by vesicle transport. Vesicles bud from their donor membrane, travel a certain distance freely or along the cytoskeleton and then fuse with a specific target membrane. Secretory vesicles can fuse with the plasma membrane and deliver their content to the extracellular space.

Finally, we discuss signal transduction mechanisms for chemical signaling molecules, which cannot pass the membrane. They are bound by transmembrane receptors, which are connected to several signal transducers. These activate secondary messengers inside the cell, which instruct the cell to react to the first message delivered at the membrane.

A. Böttger, J. Moneer, *Basic Cell Biology for Dentists and Medical Students*, https://doi.org/10.1007/978-3-032-03636-0_3

3.1 Membrane Structure

All cells are surrounded by biological membranes (biomembranes). Membranes separate unicellular organisms, such as protozoans or bacteria, from their environment. In multicellular organisms, their membranes also shield the cells from their environment within a tissue or body. Biomembranes are composed of 3–5 nm-wide lipid bilayers held together by noncovalent bonds. Proteins are incorporated into these bilayers. An early observation of lipid layers at membranes, or "a layer of fatty substances that is two molecules thick," as he called it, was made by Gorter in 1925 when he analyzed blood cells (Gorter & Grendel, 1925).

The main components of lipid bilayers are phospholipids (Fig. 3.1), (Dowhan, 1997). As described in Chap. 2, lipids are triglycerides. In a lipid, each of the three hydroxyl-groups of glycerol is engaged in an ester bond with a fatty acid. Phospholipids have only two fatty acids attached to glycerol, while the third hydroxyl group forms an ester bond

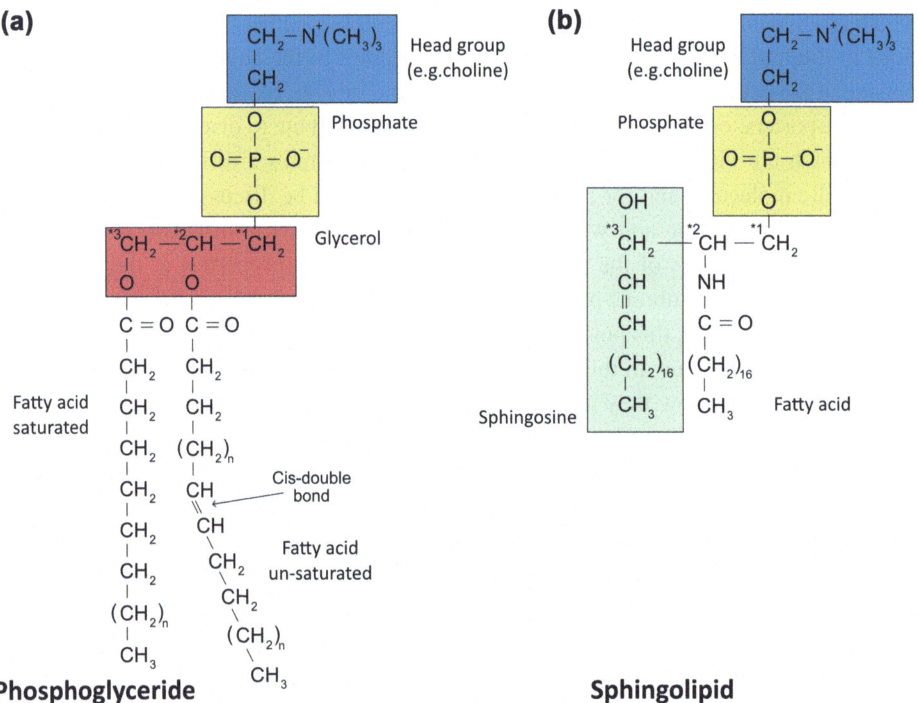

Fig. 3.1 Molecular structure of phospholipids: (**a**) Molecular structure of a phosphoglyceride (lecithin); Glycerol part is marked by red box; Phosphate yellow and head group (choline) blue; Saturated fatty acid is attached to glycerol on C3* of glycerol and unsaturated fatty acid is attached on C2*, cis-double bound is indicated; (**b**) Molecular structure of a sphingolipid (sphingomyelin), sphingosine is marked by green box, phosphate yellow, and head group (choline) blue; Sphingosine is amid-linked to acyl group of fatty acid on C2*

with phosphate. Thus, they are phosphoglycerides. With a free phosphate, the simplest phosphoglyceride is phosphatidic acid. It is a key building block of the membrane; however, it does not occur within membranes in high concentrations. All other membrane phosphoglycerides have additional, so-called head groups attached to the phosphate by a phospho–ester bond (Fig. 3.1a). The most prominent head groups are provided by small molecules including serine, ethanolamine, choline, or inositol, making the phosphoglyceride a phosphatidylserine, phosphatidylethanolamine, phosphatidylcholine (lecithin), or a phosphatidylinositol. In addition to phosphoglycerides, the group of sphingolipids also belongs to membrane phospholipids. Their backbones are made of sphingosine, a simple nonsaturated amino alcohol with a long hydrocarbon chain on its C3 atom, and an amino group on C2, which can be connected to a fatty acid. The C1 phosphate can be linked with any of the head groups mentioned above, whereby choline attachment leads to a sphingomyelin (Fig. 3.1b).

Phospholipids are nonpolar and polar at the same time. Thus, they are amphiphilic. The head group-phosphate is hydrophilic, and the fatty acids are hydrophobic. The phosphate part will interact with water, and the fatty acids will interact with lipids. Therefore, in an aqueous environment, fatty acid tails will face each other, thus forming a lipid bilayer. How pure lipids or triglycerides behave in water can easily be seen when oil is mixed with water. Lipids will quickly separate from the water. If a phospholipid is added to a beaker with water and oil swimming on the surface, its polar group will face the water, and the fatty acids will face the oil (Fig. 3.2a). Moreover, phospholipids will surround a drop of oil within a watery solution by pointing their polar groups towards the water. The fatty acid tails will assemble with the oil (Fig. 3.2b). To make it more plausible, the phospholipid could be replaced by a detergent. Similar to phospholipids, detergents have long carbohydrate chains and polar groups in their structure. As we know from washing dishes, they help to remove fat from our plates and cups. They surround the fat in the same way as the

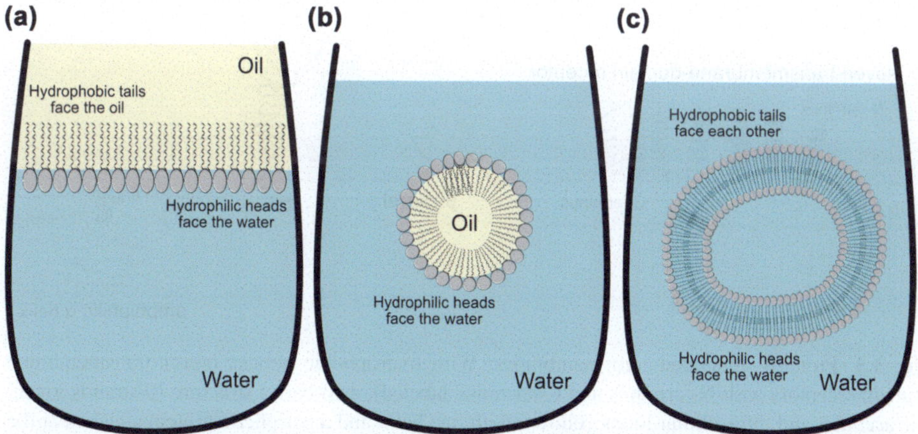

Fig. 3.2 Phospholipid arrangements: (**a**) On oil–water boundary; (**b**) With a drop of oil in aqueous solution; (**c**) In the absence of oil in aqueous solution forming a lipid bilayer

phospholipid surrounds it in our example in Fig. 3.2b. In the absence of oil, e.g., in a hydrophilic environment, phospholipids turn their hydrophobic tails toward each other and form a *lipid bilayer* (not to be confused with a double membrane, as seen in mitochondria and the nucleus). Now the polar head groups face the water on two sides, the cell exterior and the cytoplasm (Fig. 3.2c). This bilayer provides the scaffold of all biological membranes.

The plasma membrane is more than a mere physical barrier of the cell. It has many physiological functions, including exchanging compounds between the extracellular and intracellular space, maintaining an electrical potential and receiving and interpreting molecular signals to regulate cell functions in the context of the whole organism and/or in response to the environment. These functions are mediated by proteins, which are inserted into the bilayer in many different ways (Fig. 3.3). Some proteins are *integral membrane* proteins. They transverse the lipid bilayer with stretches of hydrophobic amino acids. To span the bilayer at its width of 3–4 nm, such stretches have to be ca. 20 amino acids long. They usually have a helix secondary structure, which is associated with the lipids inside the bilayer. We find many proteins that span the membrane only one time, but others do so many more times, and seven-transmembrane receptors, as the name suggests, do so seven times (Fig. 3.3). Their primary sequence therefore includes seven stretches of about 20 hydrophobic amino acids. Membrane proteins can also form a barrel structure. The outside of the barrel, which is embedded in the lipid bilayer, contains hydrophobic amino acids. Hydrophilic amino acids face the inside of the barrel (Fig. 3.3). Such proteins are very rare in the plasma membranes of animal cells; however, they are found in bacterial and also in mitochondrial membranes. A barrel structure allows the formation of membrane pores with very small diameters, which are, for instance, present in mitochondrial membranes

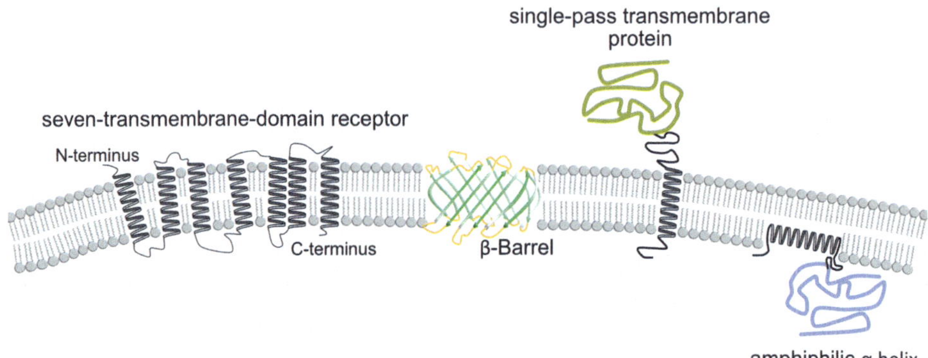

Fig. 3.3 Proteins integrated into membranes: With hydrophobic helices (seven-transmembrane domain receptors with N-terminus and C-terminus labeled); as β-barrel structure (β-strands green, extracellular and intracellular loops yellow); with one helix and a peripheric protein attached at the outside of the membrane (green); with amphiphilic helix, polar side facing the cytosol and hydrophobic part facing lipid layer, peripheric membrane protein attached at cytosolic side (blue)

and allow import of unfolded proteins. Another mechanism of protein association with membranes is insertion of an amphiphilic helix only in the inner or only in the outer lipid layer. Additionally, proteins can be post-translationally modified by attachment of fatty acids to specific cysteine residues via thioester bonds (e.g., palmitoylation by palmitic acid). These fatty acids then insert into lipid membrane sheets and thereby tether the protein to the membrane. Finally, proteins, e.g., that of the extracellular matrix, attach to the membrane indirectly by binding to integrated membrane proteins, and these are called peripheric membrane proteins. Such attachments can also be mediated by phospho–ester formation of phospholipids with sugar residues of glycosylated proteins (Fig. 3.3).

Integral membrane proteins are involved in many cell functions. They form ion channels, ion pumps, transporters, and receptors, which regulate exchange of substances and information between the cell interior and the extracellular environment. Peripheric membrane proteins form links with the cytoskeleton and proteins of the extracellular matrix. All membrane proteins participate in signal transduction.

Within the two-dimensional sheet of the phospholipid bilayer, integral membrane proteins are mobile; however, they cannot move in or out. They are inserted into the membrane already during translation and confined to membranes or membrane vesicles for all of their life, with a few exceptions as we will see later in this chapter.

All membranes have different compositions and physical characteristics precisely adjusted to functional requirements of the cell type. A major membrane component is cholesterol. Cholesterol is produced not only in liver cells but also in other organs, including the brain and the adrenal gland. Synthesis starts with AcetylCoA, the entry compound of the citrate acid cycle (see Chap. 1) and proceeds over many steps via melanonic acid and squalene, a 30 carbon nonsaturated carbon hydrate chain to the 27 carbon, heterocyclic cholesterol. This process requires ATP and $NADH_2$. Cholesterol represents the precursor for the production of steroid hormones, vitamin D, and bile acids, which are essential factors for fatty acid digestion. It was first discovered in gallstones, hence the name *chole* = Greek bile and *stereo* = Greek solid. Sterols influence the fluidity of the membrane. With a polar head group attached to a rigid steroid ring and a hydrophobic carbohydrate tail, cholesterol assembles with the fatty acids of the lipid bilayer. Cholesterol contributes to the stiffness of the membrane, and membranes with a lower content on cholesterol are more fluid (Fig. 3.4).

An early model of the structure of the membrane integrating its major membrane components comprising phospholipids, proteins, and cholesterol was provided in 1972 and called the fluid mosaic model (Singer & Nicolson, 1972). It considered the protein content of the membrane and the fluidity, which allows lateral mobility of proteins within the two dimensions of the membrane sheet. However, further research revealed a very high complexity of membrane structures with many substructures differing in the density of proteins and lipids. In this way, functional microdomains of membranes could be created (Engelman, 2005). An example consists of "lipid rafts." They represent clusters with high content of cholesterol and sphingolipids, which move within the lipid bilayer and are supposed to form functional units with specific proteins (Simons and Ikonen (1997) and a recent review by Levental et al. (2020)).

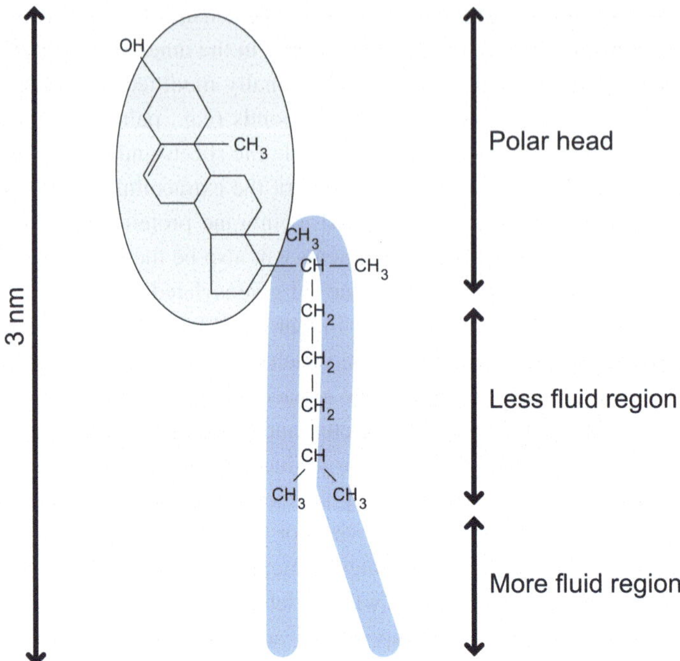

Fig. 3.4 Structural formula of cholesterol and position of polar head group and hydrophobic carbohydrate tail within the membrane, and membrane fatty acids are schematically illustrated in blue. Presence of cholesterol makes membrane stiff (less fluid), and membrane regions without cholesterol are more fluid

3.2 What Can Cross a Biological Membrane?

Cells need to exchange substances with their environment. This is regulated by the membrane permeability. Membranes are permeable for gases, such as oxygen, carbon dioxide, and nitrogen gas. They are also permeable for small hydrophobic molecules, such as steroid hormones, allowing free passive diffusion of such substances. In addition, some small uncharged, but polar molecules can pass, for instance, ethanol. Interestingly, water molecules can pass the membrane too, but only to a certain extent, above which they use specific "water channels," so-called aquaporins. For larger uncharged, but polarized molecules, like glucose and fructose and charged molecules, such as ions, amino acids, ATP, metabolites like glucose-6-phosphate, proteins, carbohydrates and nucleic acids, the membrane constitutes a complete barrier (Fig. 3.5).

How can molecules beyond free passive diffusion get in and out of cells? There are several possibilities involving specific cellular transport proteins inserted into the mem-

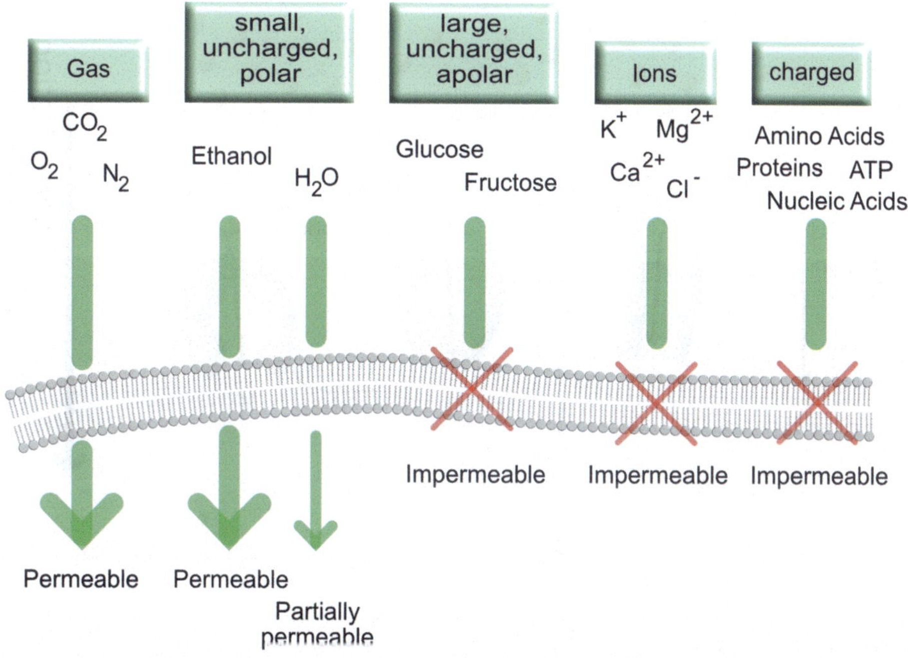

Fig. 3.5 Membrane permeability for different substances; green arrows indicate permeability or partial permeability, and red crosses indicate no permeability

brane, including channels and transporters (Fig. 3.6). Channels constitute pores with a more or less hydrophilic inner lining. They allow passive transport or eased diffusion of molecules according to their concentration gradient between the outside and inside of the cell. Substances can only pass when the channels are open. Ion channels are an example. Their openings and closings are tightly regulated, and they are "gated." "Gate keepers" include membrane voltage, extracellular ligands such as neurotransmitters, intracellular ligands, or mechanical stimuli (Fig. 3.7).

As indicated in Fig. 3.6, active transport against the concentration gradient of a certain substance requires energy. This energy can be provided by ATP. Transporters using ATP are often called "ATP-driven pumps." An example is the proton pump that fills lysosomes (see Fig. 1.1) with protons to provide the acidic pH of this cell organelle. Another source of energy arises when the transport of two substances with opposing concentration gradients between the outside and inside of the cell is coupled. In this case, substance A travels according to its concentration gradient and this transport provides the energy to transport substance B against its concentration gradient through the membrane. ATP-driven pumps

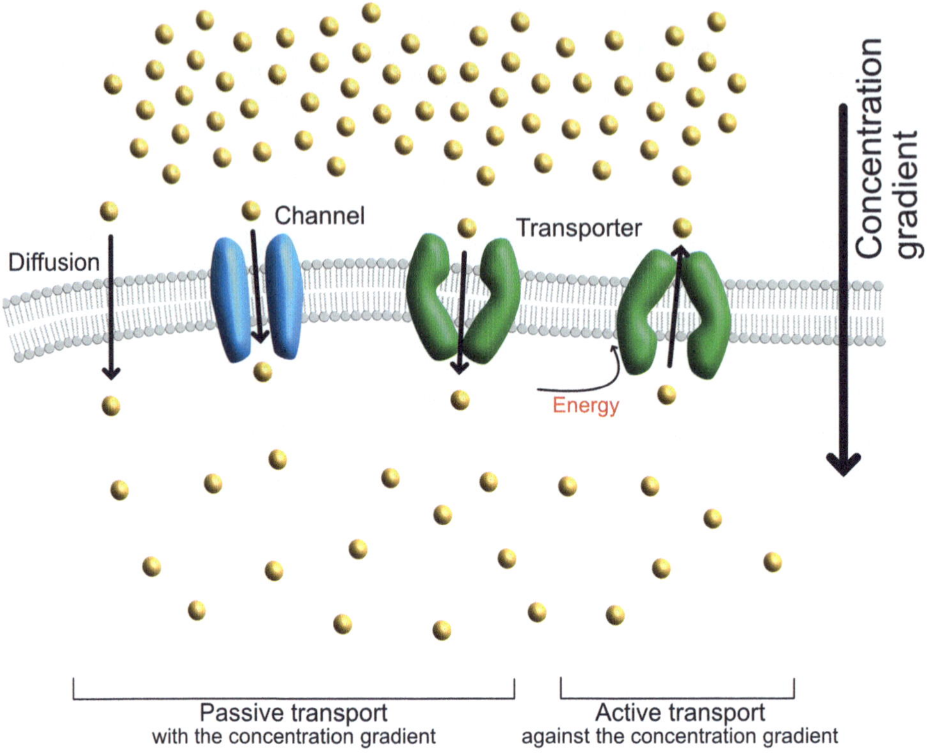

Fig. 3.6 Free diffusion, ion channels, and transporters: Concentration of substances (yellow dots) outside and inside of cell is indicated by their density; Direction of gradient is shown by black arrow; Free diffusion and diffusion through open channels work according to the concentration gradient, and transporters may transport substances with or against the concentration gradient; The latter requires energy as shown in red letters

can also be coupled with transporters and provide the energy for the latter to drive a molecule through the membrane against its concentration gradient.

The movements of ions from one side of the membrane to the other result in a disparity of concentrations of positively and negatively charged ions on both sides of the membrane. The membrane lipid bilayer constitutes an insulator (di-electricum) enabling the membrane to store electrical charges like an electric capacitor. The resulting voltage is the membrane potential. It can be measured by placing electrodes inside and outside of a cell membrane. By convention, it is measured inside versus outside and in resting cells, it values −70 mV, indicating excess negative charge at the inner side of the membrane (Fig. 3.8a). This results in an immense electric field intensity of 20×10^6 V/m (compare it with the 200 V/m of our high-voltage transmission lines!). The concentrations of the main

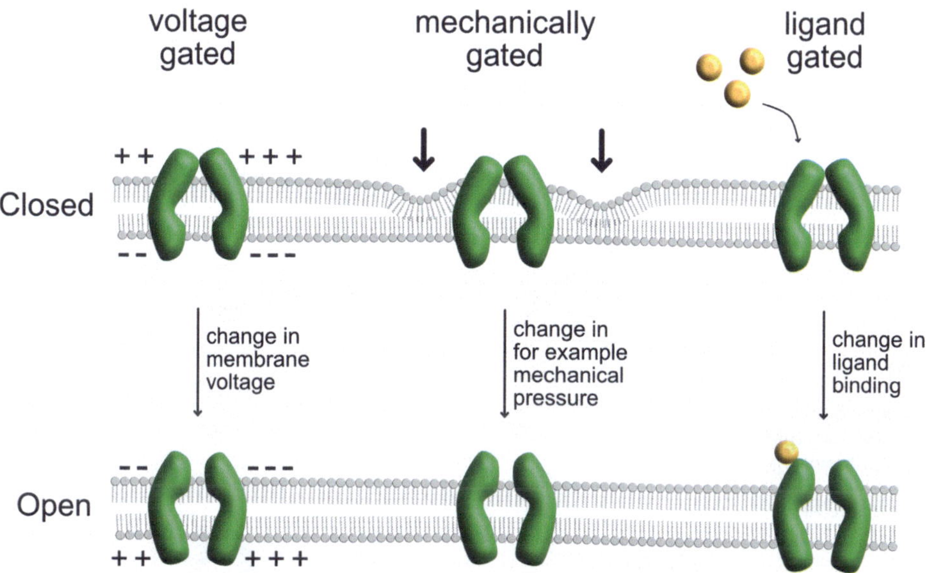

Fig. 3.7 Gated ion channels in closed conformation (upper panel) and open conformation (lower panel); Channels open by voltage change, mechanical pressure or ligand binding as indicated

cellular ions inside and outside of resting cells of all cell types are relatively similar. We find 140 mM K^+ inside cells versus 4 mM outside, 12 mM Na^+ inside versus 145 mM outside, and 0.5 μM Ca^{2+} inside versus 1800 μM outside. These concentrations are maintained by the activity of ATP-driven ion pumps. The most prominent one of those is the $3Na^+/2K^+$ ATPase. As the name suggests, it hydrolyzes one molecule ATP to provide energy for three Na^+-ions to be transported out of the cell (against the concentration gradient). This transport changes conformation of the transporter to allow binding of two K^+-ions at the outside and their transport inside and also against the concentration gradient (Skou & Esmann, 1992) (Fig. 3.8b).

Furthermore, K^+ and Na^+-channels contribute to the negative resting cell membrane potential. Some K^+-channels are always open, allowing ion exchange through the membrane along the concentration gradient. K^+ flow stops when the equilibrium diffusion potential is reached. This depends on the sum of all charges along the membrane, including the concentration of K^+ on both sides of the membrane and on some physical parameters and can be approximated according to the Nernst equation (see Box 3.1).

In contrast, Na^+ channels are mostly closed. They are "gated." In excitable cells, such as nerve and muscle cells, they may open as a result of changes in membrane voltage or neurotransmitter binding. Then Na^+-ions enter the cell, and this inward flow of positively

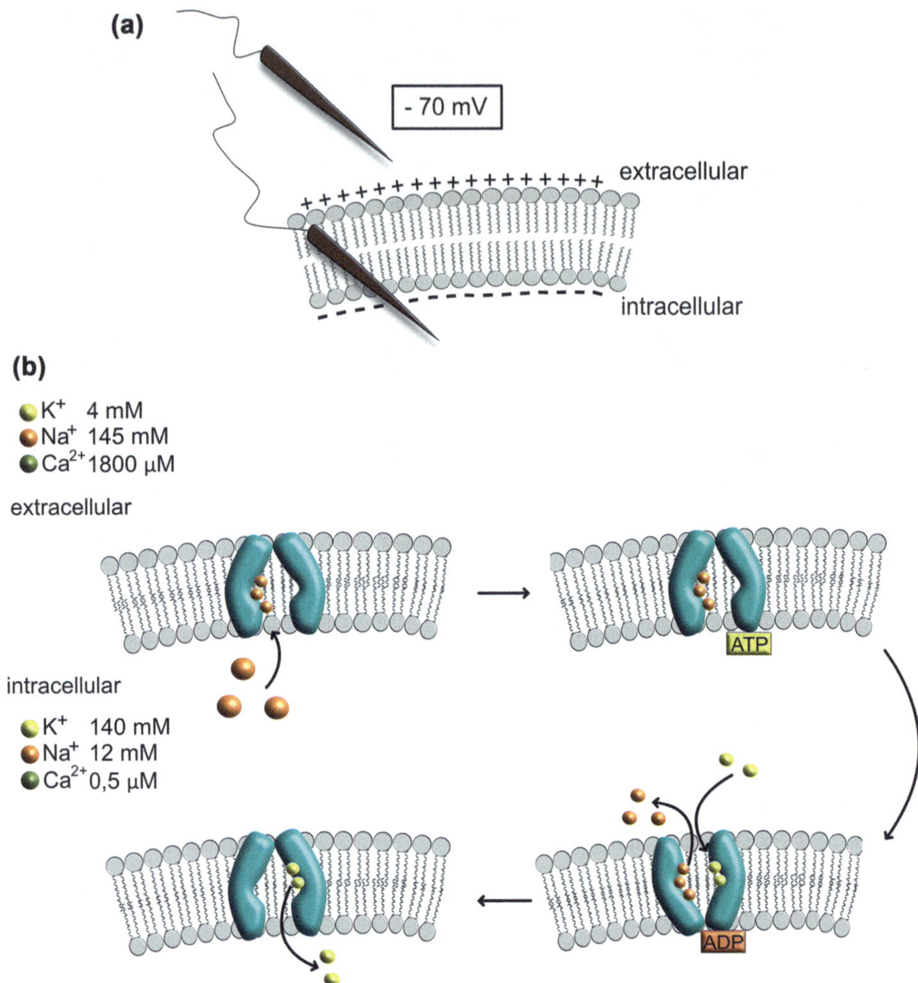

Fig. 3.8 Membrane potential and function of $3Na^+/2K^+$ ATPase: (**a**) Membrane potential measured with electrodes positioned at both sides of membrane amounts to −70 mV in resting cells as indicated; (**b**) $3Na^+/2K^+$ ATPase transports Na^+ and K^+-ions against both their concentration gradients, membrane concentrations of K^+ (yellow spheres), Na^+ (orange spheres), and Ca^{2+}-ions (green spheres) are given to indicate concentration gradients, and energy for ion transport is provided by ATP hydrolysis; ATP binding opens channel to inside, ATP hydrolysis to ADP leads precedes opening toward outside

charged Na^+ ions increases the membrane potential, resulting in depolarization (Fig. 3.9). When the membrane potential reaches +30 mV, we talk about an action potential, which can travel down an axon of a nerve cell and be transmitted to dendrites of the next nerve cell either at a chemical synapse or at an electrical synapse.

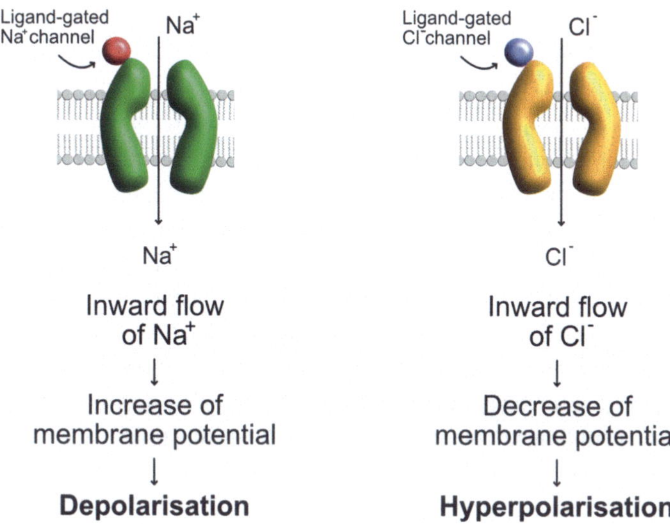

Fig. 3.9 Ligand-gated ion channels: Left-hand panel: Schematic representation of ligand (red sphere) binding Na^+-channel (green) causes Na^+ influx and increase of membrane potential (depolarization); Right-hand panel: ligand (violet sphere) binding Cl^-channel (yellow) causes Cl^- influx and decrease of membrane potential (hyperpolarization)

Box 3.1 Nernst Equation

The Nernst equation describes the establishment of electrochemical potentials by diffusion of the anions and cations of a salt. When a salt is located at the bottom of a water filled beaker, it diffuses from its place of high concentration into the water. The velocity of diffusion depends on the size of the individual particles in the solution and their concentration. Salts will divide in solution into their positively charged cations and negatively charged anions (e.g., K^+ and Cl^- if the salt is KCl). The smaller ion moves faster than the larger one and diffuses ahead of the latter. This creates a gradient, which is limited by the electrical attraction of the cations and anions. However, in a certain area of the solution, a difference in the electrical standard potential persists. The Nernst equation defines the size of this potential difference (E) in relation to the standard potential (E_0), which can be called diffusion potential, and as such, it can be applied to the diffusion of K^+-ions through open ion channels. The Nernst equation includes the following parameters: ionic valence (Ionenwertigkeit) (z), mobility of the cation [u^+] mobility of the anion [u^-], ion concentration of the salt on the concentrated side (a_2) and on the diluted side (a_1), the gas constant R, the absolute temperature T and the electrostatic power (F = Faraday constant = 96,500 C/g-eq). The Nernst equation then looks as follows:

$$E = E_0 - \frac{RT}{zF} \times \frac{u^+ - u^-}{u^+ + u^-} \times \ln \frac{[a_2]}{[a_1]}$$

A decrease in membrane potential, maybe due to influx of negatively charged Cl^- ions, is called hyperpolarization (Fig. 3.9). At chemical synapses, the neurotransmitters acetylcholine and glutamate, for instance, open cation channels (Na^+, Ca^{2+}), leading to depolarization and activation of neuronal electrical transmission. Acetylcholine-gated ion channels are called nicotinic acetylcholine receptors because nicotine activates such receptors. Nicotinic acetylcholine receptors are also present at all neuromuscular junctions where they open in response to acetylcholine secreted from motor neurons to activate muscle contraction. The neurotransmitters glycine and GABA open chloride channels, leading to hyperpolarization and inhibition of neuronal electrical transmission. This activity can be mimicked by drugs such as barbiturates or benzamidines that therefore have calming properties. Electrical synapses are found in heart muscle cells and in the brain. They constitute cell junctions, forming channels for ions to pass from cell to cell. This allows a direct propagation of membrane potential changes (GAP-junctions, see later) (Böttger et al., 2018).

3.3 Membrane Biosynthesis

As we have seen, the membrane consists of phospholipids and proteins. Phospholipids are synthesized at the smooth *endoplasmic reticulum* (ER). Proteins are synthesized on ribosomes in the cytoplasm. The question is: how do proteins get integrated into membranes and how do proteins get out of cells, for instance, into the extracellular space to form an extracellular matrix?

Ribosomes are found in two locations in cells. Many are freely distributed in the cytoplasm. However, ribosomes are also found tightly associated with membranes of the ER in the so-called rough ER (see Chap. 1) (Chua et al., 1976). These are the ribosomes where extracellular or membrane proteins are translated. Cytoplasmic and ER-associated ribosomes are basically biochemically indistinguishable. How is the mRNA encoding a membrane protein recognized and targeted to ER ribosomes? This is a sophisticated process. In fact, any mRNA leaving the nucleus is targeted to cytoplasmic ribosomes to start translation. For ER targeting, the first translated amino acids leaving the ribosome are crucial. If they constitute a specific sequence, called a "signal peptide", the ribosome will be transported to the ER membrane and the protein chain will be inserted into the ER *co-translationally*, meaning while it is translated (Gilmore et al., 1982). How this works is illustrated in Fig. 3.10a and described next.

The signal peptide that has been translated and is coming out of the ribosome is recognized by a specific protein–RNA complex, the "signal recognition particle" (SRP). This particle "grabs" the ribosome by the emerging signal peptide, stopping translation in the process, and takes it directly to an SRP receptor on the ER membrane. From the SRP receptor, the ribosome is transferred to a channel in the ER membrane, the translocon. The ribosome is put onto the translocon in such a way that the nascent protein chain will directly be guided through the channel of the translocon into the lumen of the ER. Translation is taken up again after the SRP has been removed. The signal peptide, when entering the translocon, is cleaved off (Fig. 3.10a). Therefore, proteins found to be secreted are usually

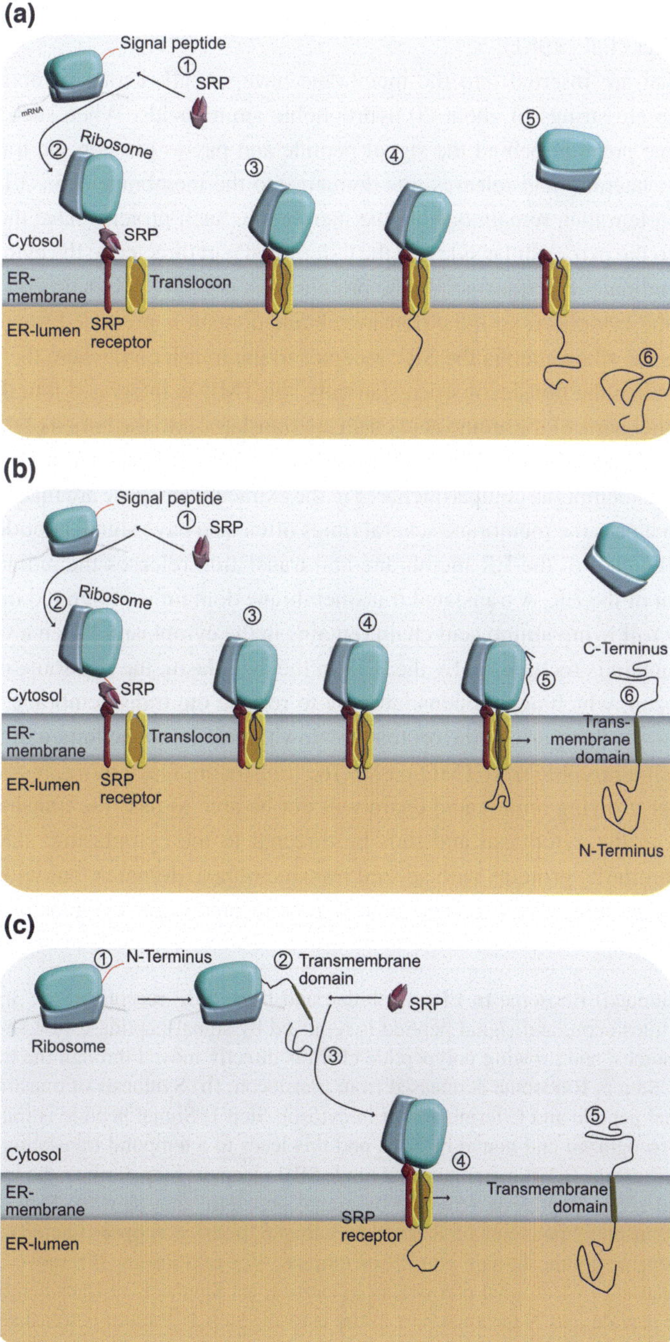

Fig. 3.10 Co-translational import of proteins into ER lumen and into ER membranes: (**a**) secretory proteins with N-terminal signal peptide: Step 1: Signal peptide is translated by the ribosome and recognized and bound by SRP, and this leads to a temporal translation stop; Step 2: SRP transfers

shorter than expected by the length of the predicted amino acid chains (Walter & Blobel, 1981a, b; Walter et al., 1981).

Proteins that are inserted into the membrane have so-called transmembrane domains (TMD). These are strings of about 20 hydrophobic amino acids. When such a domain is present at some position behind the signal peptide and passes through the translocon, the channel opens laterally and releases this domain into the membrane (Fig. 3.10b). Further parts of the protein then remain outside the membrane. Such proteins have the N-terminal part directed to the extracellular side and the C-terminal part directed to the cytoplasmic side of the ER membrane. If a transmembrane protein does not have a signal peptide, it will be translated in the cytoplasm until the transmembrane domain is presented to an SRP. Then, the SRP takes the ribosome via the SRP receptor to the translocon, where the TMD enters the channel. When the translocon opens laterally, the TMD is integrated into the ER membrane and further parts of the amino acid chain are translated into the lumen of the ER. In this way, transmembrane proteins with an N-terminal cytoplasmic part and a C-terminal part localized within membrane compartments or in the extracellular space are made (Fig. 3.10c).

Proteins that pass the membrane several times often also have signal peptides. Thus, the ribosome is directed to the ER membrane and translation releases the amino acid chain into the lumen of the ER. A translated transmembrane domain is integrated into the membrane and the following amino acid chain remains in the cytoplasm. When a second transmembrane domain is recognized by the SRP in the cytoplasm, the ribosome is taken back to the translocon, which again opens laterally to release the transmembrane domain into the membrane. Further parts of the protein are now targeted into the lumen of the ER. This can go on until a possible next TMD passes the translocon. Once more, it opens laterally and the further growing amino acid chain will not be able to pass the translocon but will be translated in the cytoplasm and thus be directed to the cytoplasmic side of the ER membrane. Similarly, proteins with several transmembrane domains, but without a signal peptide, are made; now the N-terminal part will be located in the cytoplasm.

Fig. 3.10 (continued) ribosome to ER membrane and binds SRP-receptor; Step 3: Ribosome is transferred to translocon, and signal peptide is removed by signal peptidase (not shown); Step 4: Translation resumes, and growing polypeptide chain is directly moved through the translocon into the ER lumen; Step 5: Ribosome is released from translocon; (**b**) Synthesis of transmembrane proteins with signal peptide and C-terminal part in cytosol: Step 1: Signal peptide is translated by the ribosome and recognized and bound by SRP, and this leads to a temporal translation stop; Step 2: SRP transfers ribosome to ER membrane and binds SRP-receptor; Step 3: ribosome is transferred to translocon, and signal peptide is removed by signal peptidase (not shown); Step 4: Translation resumes until transmembrane domain is translated; Step 5: translocon opens laterally (black arrow) and releases transmembrane domain into ER membrane; Step 6: ribosome is removed from translocon, and translation of C-terminal part occurs in cytosol; (**c**) Synthesis of transmembrane proteins without signal peptide and N-terminal part in the cytosol; Step 1: Protein is translated until transmembrane domain appears in cytosol; Step 2: SRP binds transmembrane domain and translation stops temporarily; Step 3: SRP transfers ribosome to ER membrane and binds SRP receptor (not shown), and ribosome is transferred to translocon; Step 4: Translocon opens laterally to release transmembrane domain into the membrane (black arrow); Step 5: Translation resumes, and growing C-terminal part of polypeptide chain is directly moved through the translocon into the ER lumen

After being co-translationally translocated into the ER as described above, proteins are folded by specific ER-resident chaperones (see Fig. 2.22). The ER is the first compartment for the "secretory pathway." From the ER, proteins travel in vesicles via the Golgi apparatus to lysosomes and endosomes, back to the ER, to secretory vesicles, or directly out of the cell. Each protein has a specific final destination and carries sorting signals, some of which are applied in the Golgi apparatus by post-translational modification, e.g., glycosylation (Blobel, 2000). All proteins that enter the secretory pathway once, remain within the membrane system during their lifetime. They will be part of membrane compartments, either as membrane or as luminal proteins. They may end up inserted into the plasma membrane or will be secreted into the extracellular space (Fig. 3.11). They travel between

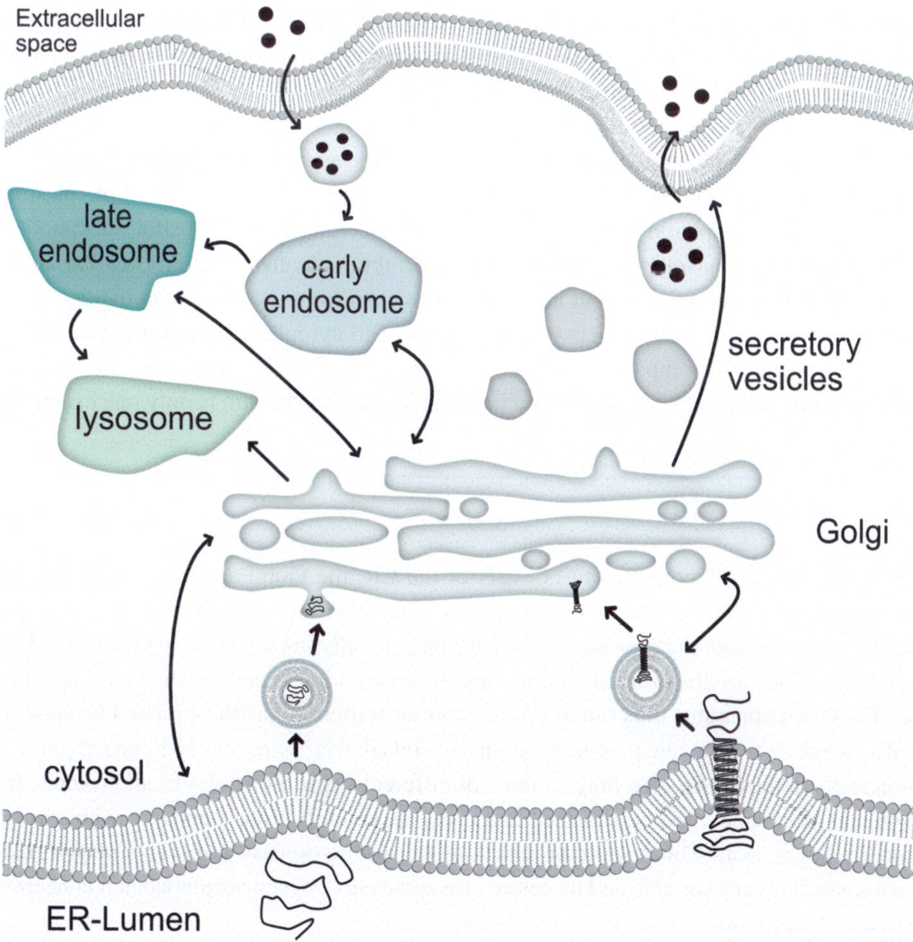

Fig. 3.11 Compartments of the secretory pathway and vesicle trafficking: Vesicles bud from ER and fuse with Golgi membranes and vice versa, vesicles budding from the trans-Golgi fuse with ER membranes, vesicles budding from the Golgi travel via endosomes to lysosomes or to the plasma membrane. Vice versa, vesicles from the plasma membrane fuse with endosomal compartments and from there travel to lysosomes via endosomal storage compartments (early endosomes, late endosomes, and recycling endosomes (not shown)) or to Golgi

membrane compartments in vesicles, which bud off from one membrane and fuse with another. This works in two directions. After synthesis proteins get transported into ER exit compartments, bud off from the ER, fuse with and bud off the Golgi cisternae, fuse with lysosomes, or grow into secretory vesicles, which fuse with the plasma membrane and empty their content into the extracellular space. By endocytosis, however, extracellular material or membrane receptors can also bud off in endosomal vesicles from the plasma membrane into the cell. These vesicles can then fuse with lysosomes for degradation or remain in intermediate compartments, from which they may return to the plasma membrane. ER-resident proteins usually also leave the ER and enter the Golgi to get glycosylated or processed. They then bud off from the Golgi and return into the ER to carry out their jobs (Fig. 3.11). Exceptions from the "once in the membrane—always in a membrane" rule are misfolded proteins. These are recognized by the ER quality control system and can be translocated back into the cytoplasm, where they are degraded in proteasomes (Balch & Rothman, 1985; Scott & Schekman, 2008).

Membranes are asymmetric with respect to the side facing the cytoplasm and the side facing the extracellular space. This is partially due to the asymmetric distribution of phospholipids. Sphingomyelins are largely found in the outer membrane leaflet of the plasma membrane, and phosphatidylserine, phosphatidylethanolamine, and inositol phospholipids are almost exclusively found in the inner membrane leaflet (Fujimoto & Parmryd, 2016). In addition, protein content at the outside differs from the inside and extracellular proteins are usually heavily glycosylated in contrast to cytoplasmic proteins. The redox state in the secretory compartments is more oxidized than in the cytoplasm and accordingly disulfide bridges are usually found in secreted proteins. Insulin is only one example.

3.4 Vesicle Traffic

The entire secretory compartment consists of the ER, the Golgi apparatus, endosomes, lysosomes, secretory vesicles, and the plasma membrane. As described above, proteins present in the secretory compartment are co-translationally inserted into the ER or the ER membrane. They are then transported to their destination in the cell via the Golgi apparatus. The Golgi apparatus thus functions as a sorting station. It has to be passed because it is also the place for specific post-translational O-linked glycosylations. N-linked glycosylation starts in the ER by attaching a "tree" of different sugar molecules to all proteins. In the Golgi, this "tree" will be trimmed and plugged, so each protein ends up with a very specific "sugar coat." This sugar coat serves in the sorting process, e.g., for transport into lysosomes. It is very variable and therefore also serves in cell recognition when it concerns extracellular proteins.

All transport within the secretory pathway is vesicle transport. Vesicles bud from one compartment and fuse with the next. Secreted proteins are stored inside secretory vesicles as long as they are in the cell. These vesicles then fuse with the plasma membrane and release their content during secretion. This can happen in a constitutive manner. Proteins of the extracellular matrix, such as collagen, are constantly produced and secreted into the extracellular space. However, many substances are only secreted "on demand". These are

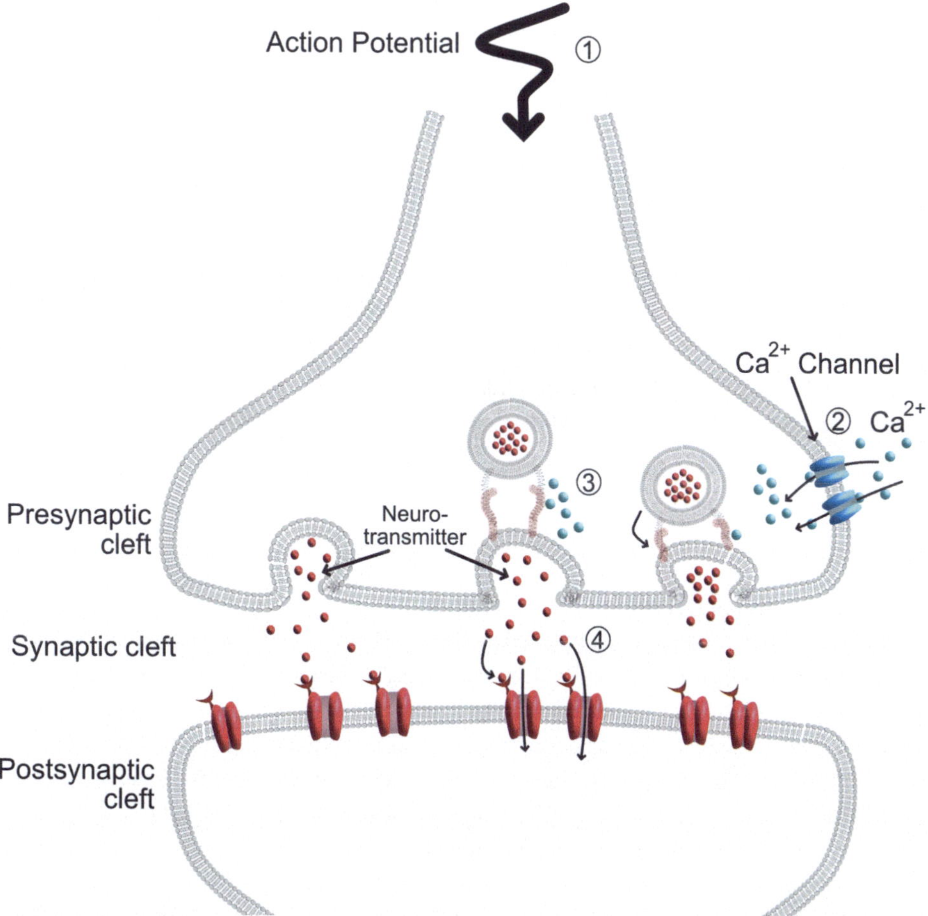

Fig. 3.12 Ca^{2+}regulates neurotransmission release at chemical synapses: Step 1: Axon potential arrives at synapse; Step 2: Voltage-gated Ca^{2+}channels open; Step 3: Ca^{2+}reaches membrane-tethered neurotransmitter-filled vesicles; Step 4: vesicles release neurotransmitter into the synaptic cleft

stored in vesicles and are released in response to certain stimuli. An extreme case are neurotransmitters stored in secretory vesicles close to the presynaptic membrane. They are quickly secreted when an action potential arrives and voltage-dependent Ca^{2+}channels are opened. Such Ca^{2+}-influx, for instance, allows instant neurotransmitter release into the synaptic cleft (Fig. 3.12) (Rothman, 2002).

Another type of secretion involves the release of whole vesicles into the extracellular space. Such vesicles are called "matrix vesicles" and they are tethered to the extracellular matrix after release from the cell. This release happens by outside budding from the plasma membrane (Fig. 3.13). Inside the cell when they are forming they sometimes constitute multivesicular aggregates. They seem to be especially abundant in mineralizing tissues, such as teeth and bones, where they have been originally discovered. They contain Ca^{2+} and phosphate ions in high concentrations, whereby Ca^{2+} is bound directly to phospholipids, e.g., to phosphatidyl serine, see Box 3.2.

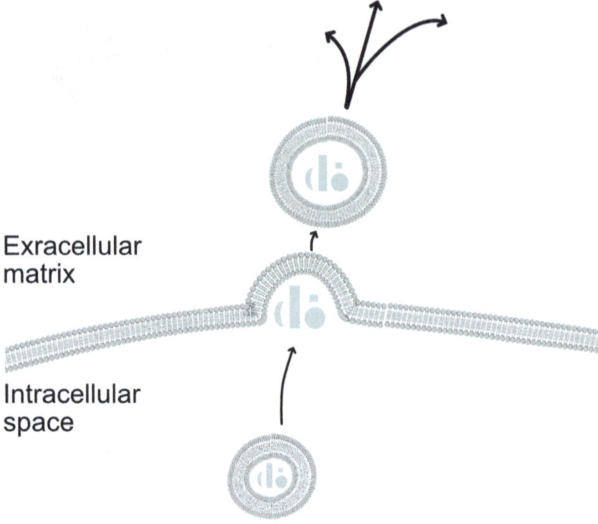

Fig. 3.13 Release of matrix vesicles into the extracellular space for producing dentin and bone for example

Box 3.2 Biomineralization

Biological mineralization is a physicochemical process where anorganic material is deposited within or outside cells. It is used quite widespread by living organisms. For instance, diatoms deposit silicate inside the cell. Magnetobacteria deposit FeO inside. Outside cells, large-scale biomineralization is found in mussels and corals (coral riffs); here calciumcarbonate is used. More interesting for dentists and medical students is calciumphosphate (apatite in nature, ($Ca_{10}(PO_4)_6(OH)_2$)), from which bones and teeth are made. Biological apatite has a different stoichiometry when compared with environmental apatite and the naturally occurring ions are replaced by other ions, e.g., Na^+, Mg^{2+}, K^+ for Ca^{2+}, $CO_3{}^{2-}$ for phosphate ions, and fluorid ions for hydroxyl ions, allowing the observed biological diversity in such material of living beings.

3.5 Vesicle Budding from Donor Membranes and Fusion with Target Membranes

Budding of vesicles from membranes is driven by polymerization of a soluble, so-called "*coat*" protein. This "coat" surrounds the budding vesicles from the beginning of the budding process until their interaction with motor proteins. Motor proteins transport the vesicles over longer distances along microtubules (see Chap. 4) towards the target membrane.

The function of coat proteins is twofold. First, coat proteins show specificity for the membrane proteins, which are to be transported within the vesicles they cover, and second, in the process of vesicle coating, the vesicle will be bent and increasingly curved to allow its budding of the donor membrane. This requires energy which is supplied by GTP-hydrolysis. Either the cytosolic part of membrane proteins interacts directly with coat proteins, or in case of soluble cargo, the luminal protein interacts with an adaptor in the membrane that is recognized by the coat. In this way, the polymerizing coat controls the content of the vesicles.

Coated vesicles are transported in anterior direction from the ER to the Golgi and into downstream compartments. Retrograde transport also uses coats, for instance, for budding of vesicles from the plasma membrane and transport into endosomal compartments and budding of vesicles traveling from the Golgi apparatus back to the ER.

For discovering the molecular mechanisms of vesicle transport, the Nobel Prize for medicine and physiology was awarded to James Rothman, Randy Shekman, and Thomas Südhof in 2013. A major component of their elucidations about the machinery for vesicle transport were coat proteins. Those are specific for the compartment from which the coated vesicles bud off, and for the traveling directions of such vesicles. For example, *Co*at *p*rotein *II* (*COPII*) is used for budding from the ER and traveling to the Golgi, and *C*oat *p*rotein *I* (*COPI*) is used for vesicles budding from the Golgi and heading back to the ER (Fig. 3.14b). For budding from the plasma membrane into the cell, Clathrin is used (Ungewickell & Branton, 1981). This process is important for recycling of membrane proteins, including receptors, reuptake of neurotransmitters into presynaptic cells and many other endocytotic processes, e.g., uptake of nutrition and pathogens (Fig. 3.14a). Moreover, the membrane of most cells permanently ingests portions of itself. During this process, liquids from the cellular environment, macromolecules, and even particles are taken up by endocytosis. Depending on the nature of this content, it can be distinguished between "pinocytosis" or "cell drinking" for liquids, and "phagocytosis" or "cell eating" for solids. For both processes, clathrin coats are formed around invaginations of the membrane (clathrin-coated pits, Fig. 3.14a), which will be constricted from the membrane into vesicles. Such vesicles quickly shed of the clathrin coat and fuse with endosomes, thereby delivering the vesicle content. Endosomes consist of different parts. The first part, which receives vesicles from the plasma membrane, is the "early" endosome. It delivers its content to the Golgi, to "late" endosomes or to "recycling" endosomes, which can bring back material to the membrane. This pathway is used for recycling receptor molecules. In order to guarantee the integrity and size of the plasma membrane, vesicles with fresh membrane material from the Golgi apparatus constantly fuse with the membrane and replenish it.

Many components of the secretory pathway have been discovered in budding yeast. Randy Schekman's group in Berkley, California, developed screening procedures in budding yeast to identify mutants defective in secretion pathways (Schekman, 2010). Because such mutants were expected to be lethal, they used "temperature-sensitive" mutants, which would behave normally at the usual temperature (so-called permissive temperature) and misfunctioned at a higher temperature only. These researchers were able to accumulate a whole set of mutants defective in secretion and the responsible genes were called *SEC* genes. The coat protein subunits forming the Coatamer II complex therefore are also named Sec23 and Sec24. These proteins are highly conserved between yeast and humans

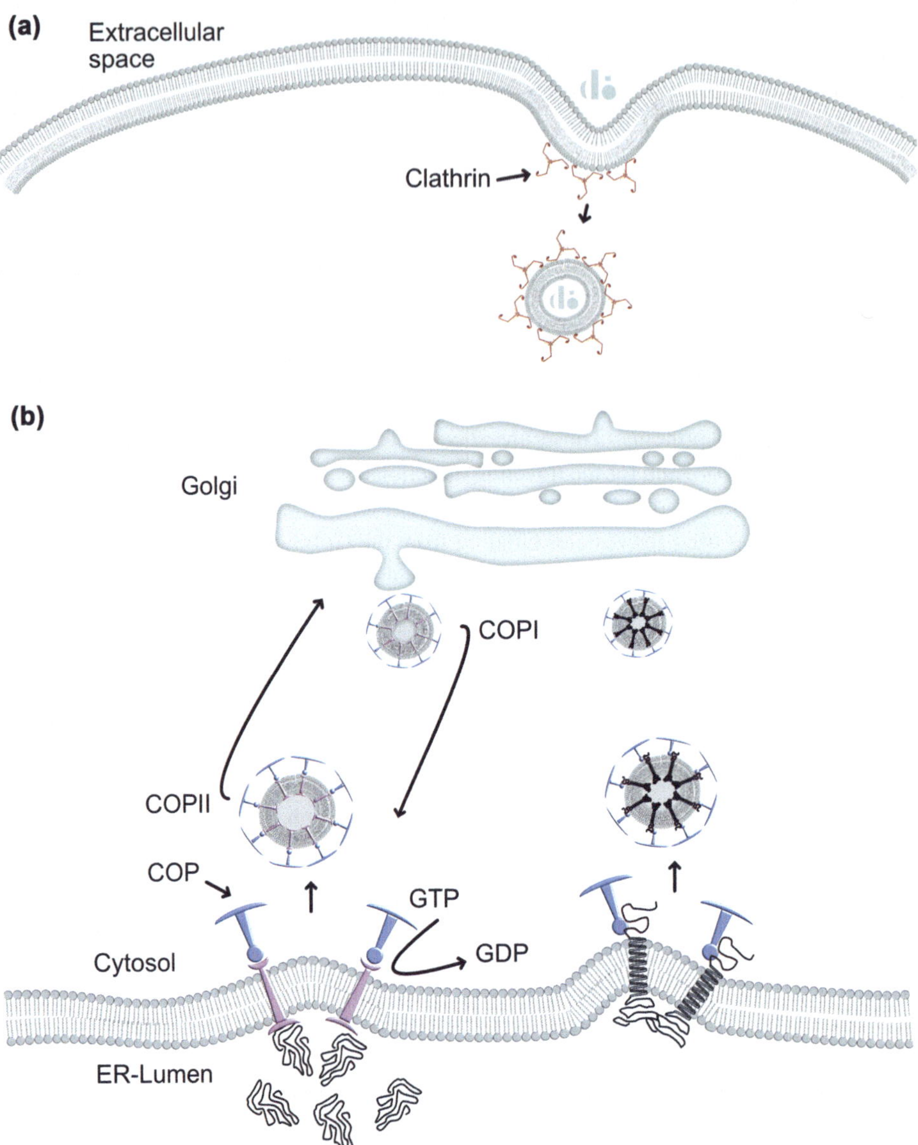

Fig. 3.14 Vesicle trafficking using coat proteins: (**a**) Schematic representation of clathrin- (red) coated membrane pit filling with extracellular material and clathrin-coated vesicle after constriction from the membrane; note that coated vesicles are short lived, shedding their coat immediately after constriction; (**b**) Schematic representation of Golgi with COPI-coated vesicles traveling from Golgi to ER and COPII-coated vesicles traveling from ER to Golgi (coat proteins blue symbols), soluble content of membrane compartments is bound by adaptors within membrane (pink symbol), which attract coat protein (blue symbol), and cytoplasmic domains of transmembrane proteins of membrane compartments bind directly to coat proteins

to the extent that a defective yeast protein can be replaced by the human version to completely recover the function of the secretory machinery in the yeast. Many of the mammalian proteins were discovered by the lab of James Rothman and eventually, the yeast and mammalian lines of research collaborated in revealing the molecular mechanisms of vesicle trafficking within an elaborated network of intracellular membranes (Schekman, 2002). In the following, as an example, *COPII* coating and uncoating are explained in more detail. One COPII complex is a dimer consisting of coat protein 1 and coat protein 2 (Sec23/24). Coat protein 1 recognizes the extracellular domain of a membrane cargo protein, whereas coat protein 2 is attached to a small GTPase switch protein Sar1 (see Box 3.3). This attachment only works when GTP is bound to Sar1. The necessary factor for activating the G protein by exchanging its tightly bound GDP for GTP (the GEF or guanyl nucleotide exchange factor) is located in the budding membrane (Sec 12). It removes GDP from Sar 1, allowing GTP to bind (see G-protein cycle as illustrated in Box 3.3). GTPSar1 then changes conformation and inserts into the membrane too, using a small hydrophobic tail that is only exposed when GTP is bound to Sar1. Sar 1 then recruits COPII coat protein 2 and membrane bending is initiated.

After budding, the coat is removed. This is induced by GTP hydrolysis of Sar1GTP. Sar1, as all small G-proteins, has an almost neglectable intrinsic GTP hydrolyzing capacity. Therefore, a GTPase activating protein is recruited for "uncoating" of the vesicle. The GTP is hydrolyzed and GDP-bound Sar 1 leaves the vesicle, inducing disassembly of the coat. The uncoated vesicle then moves along microtubules to fuse with its target membrane.

Box 3.3: GTPase Switches

GTPase switches are small G-proteins. G stands for GTPase, meaning that these enzymes can hydrolyze GTP. In their inactive state, they are bound to GDP very tightly. For activation GDP has to be replaced by GTP. This step needs to be catalyzed by a guanyl-nucleotide exchange factor (GEF) because of the close interaction of GDP with the G-protein. After removal of GDP, GTP-molecules can bind as their cellular concentration is very high in comparison to GDP. With GTP bound, the conformation of the G-protein changes and it can then perform certain tasks, such as interacting with another protein. G-proteins do have an intrinsic GTPase activity; however, this is very inefficient on its own. Therefore, GTPase-activating proteins (GAP) are required which help hydrolyzing the bound GTP to yield an inactive GDP-bound G-protein. For Sar 1, the cycle is illustrated as follows:

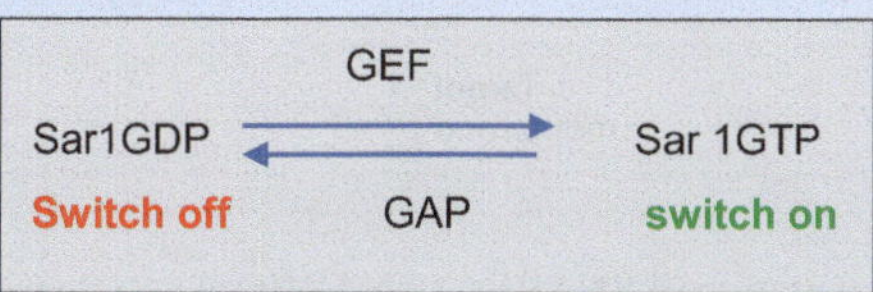

Vesicle fucion is carried out without coat proteins. The major players in this process are "*s*oluble *N*SF sensitive-factor *a*ttachment *r*eceptors," *SNAREs*. This name is derived from experimental protocols that lead to the discovery of SNARE molecules. Now the term SNARE is very familiar to researchers in the field and students and teachers alike, making it a generally accepted term, which will be used in the following. SNAREs form large complexes to allow vesicle docking to target membranes. They consist of two subgroup SNARE proteins. One is attached to the vesicle (*v*esicle-SNARE, vSNARE), and the other to the target membrane (*t*arget-SNARE, tSNARE). Examples for SNARE proteins are Synaptobrevins, Syntaxins, Endobrevin, and the name giving vesicle fusion protein NSF.

SNAREs are integrated in vesicle and target membranes with long α–helices reaching out of the vesicle into the cytoplasm (Fig. 3.15). When the vesicle and the target membranes come together, these helices on opposite membranes intertwine, providing the force for vesicle fusion. Energy is provided by small G-proteins of the rab family switching between GTP and GDP-bound states. Interaction of Rab-GTP with rab-effector proteins at the vesicle membrane is the preposition for intertwining of the helices of vSNAREs and corresponding tSNAREs leading to vesicle fusion. This mechanism is also used for exocytosis of vesicle content of secretory vesicles.

For a regulated exocytosis such as the release of neurotransmitters into the synaptic cleft in response to action potentials, a stimulus is needed to trigger the vesicle fusion with the target membrane. The trigger for vesicle fusion and neurotransmitter release is a change in the concentration of Ca^{2+}-ions in the vicinity of the vesicle. For relaying action potentials between excitable cells, Ca^{2+}-entrance into the presynaptic cell is mediated by opening of voltage-gated Ca^{2+}-channels. In the absence of such Ca^{2+}-signals, synaptic vesicles are placed close to the synaptic membrane by the v- and t-SNAREs. Intertwining of the SNARES and membrane fusion only start upon Ca^{2+}-influx after opening of voltage-gated Ca^{2+}-channels (Figs. 3.12 and 3.15) (Südhof et al., 1993).

SNAREs can be victims of bacterial toxins, e.g., botulinus toxin and tetanus toxin. These toxins represent multifunctional trimeric protein complexes, part of which are proteases targeting SNAREs in certain cells. Cell specificity is mediated by the lectin subunits, which bind to surface carbohydrates on selected nerve cells. The third subunit allows

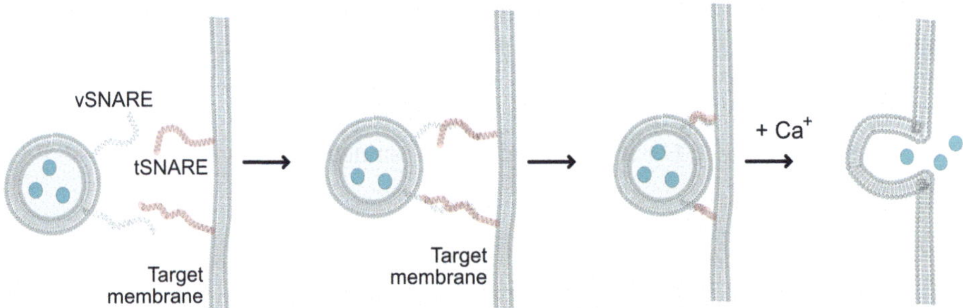

Fig. 3.15 Schematic representation of SNAREs: vSNAREs on vesicle membranes (gray) and target compartment membrane tSNAREs (red) interact and tether vesicle to target membrane; Ca^{2+}-signal arrives, vSNAREs and tSNAREs intertwine strongly, vesicle membrane fuses with target membrane, and vesicle content (blue spheres) is released

the toxin to enter the cell. Botulinus and Tetanus toxins act on synaptic SNAREs present at the neuromuscular junction. They block synaptic transmission by degrading SNAREs and thus preventing vesicle excocytosis of neurotransmitters. Botulinus toxin acts directly at the neuromuscular junction and blocks release of acetylcholine. This leads to muscle paralysis, which is deliberately induced by Botox by the cosmetic industry. However, it is also responsible for death by Botulinus poisoning due to ingestion of spoilt food. Tetanus toxin has the opposite effect on muscle contractions, and it leads to cramping. This toxin acts on an inhibitory synapse and destroys SNAREs responsible for release of the neurotransmitter glycine, which is responsible for a feedback inhibition mechanism during muscle contraction (Böttger et al., 2018).

3.6 Protein Sorting

All proteins are synthesized on ribosomes in the cytoplasm. However, in an eukaryotic cell, proteins function in every cell organelle. We have already seen that proteins of the secretory pathway reach their destination after being co-translationally inserted into the ER of the ER membrane by vesicle trafficking. The vesicles control their cargo and deliver it to the correct places within the cell. This sorting process is possible because proteins carry address stickers to indicate where they have to go. These are specific "sorting signals" encoded in the amino acid sequences of the proteins and defining their journey's end. Proteins that are integrated in the membrane display such signals on the cytoplasmic part of their amino acid chains, where they are directly recognized by coat proteins and by receptors on the target membranes. Soluble cargo residing within the lumen of the vesicle binds to receptors on the vesicle membrane, which then display sorting sequences to the coat proteins. The amino acid chain "Lys-Asp-Glu-Leu" (*KDEL* in single letter coding), for instance, flags ER-resident proteins in this way. For lysosomal proteins, a mannose-6-phosphate signal ensures vesicle transport into lysosomes.

What about mitochondrial proteins? Only a small fraction of mitochondrial proteins is translated on mitochondrial ribosomes. All the other mitochondrial proteins are encoded by nuclear DNA, translated in the cytoplasm, and then transported into the mitochondria. Mitochondrial membranes possess pore-forming proteins, the porins. The hydrophobic transmembrane parts of these molecules have a cylindrical arrangement of β-strands looking like a barrel. Inside the barrel is a water-filled cavity, which allows passing of compounds soluble in water. To the outside, hydrophobic and apolar amino acids interact with the lipophilic parts of the inner membrane. Porins allow import of unfolded protein chains. Mitochondrial proteins also have signal sequences, usually within the N-terminal sequence. Through these short amino acid chains, specific heat shock proteins are recruited. By binding to the unfolded protein, instead of initiating folding, these heat shock proteins prevent the protein chain to adapt its final 3-dimensional structure. Then they target the protein to mitochondrial pore proteins. These Porins are part of two elaborate protein complexes, the "*t*ranslocon of *o*uter *m*embrane" (*TOM*) and the *t*ranslocon of *i*nner *m*embrane (*TIM*). Only when the proteins have passed the pore, the heat shock proteins function as chaperones and help in folding the mitochondrial proteins into their functional structures. Some proteins remain as residents in the intermembrane space between the outer and the inner mitochondrial membranes. Others are inserted into either of the two membranes and some travel through both, the TOM and TIM complexes into the mitochondrial matrix. Import of proteins into chloroplasts is organized in plant cells in very similar ways.

Protein translocation into the nucleus uses nuclear pores. These "perforate" the nucleus in a very dense pattern. Two nuclear pores are about 100 nm apart and they form channels with a diameter of 30 nm. For this pore density and size, not to make the nucleus a totally permissive sieve, the very complex structure of nuclear pores prevents the passage of molecules above a certain size. The structural and functional components of nuclear pores have been analyzed in great detail by biochemical and structural investigations. Each nuclear pore complex in yeast nuclei consists of 456 single protein molecules of 30 different types. With membrane spanning proteins, they form a scaffold, where linker proteins and pore proteins are attached to (Lim et al., 2008). Nuclear pore proteins have long unstructured amino acid chains made of repeats of phenylalanine and glycine (FG), which clog the pores like woolen threads might block a sink (Fig. 3.16). Different from the sink, nuclear proteins can be transported through this mesh inside the nuclear pores when they are bound to specific receptor-like molecules called importins. To find those importin molecules in the cytoplasm, nuclear proteins usually possess "nuclear localization signals" (NLS) somewhere within their amino acid chain. The amino acid sequence P-P-K-K-K-R-K-V, for instance, constitutes an NLS. Replacing the second lysine by a threonine residue already prevents nuclear import, as was shown in 1984 (Kalderon et al., 1984).

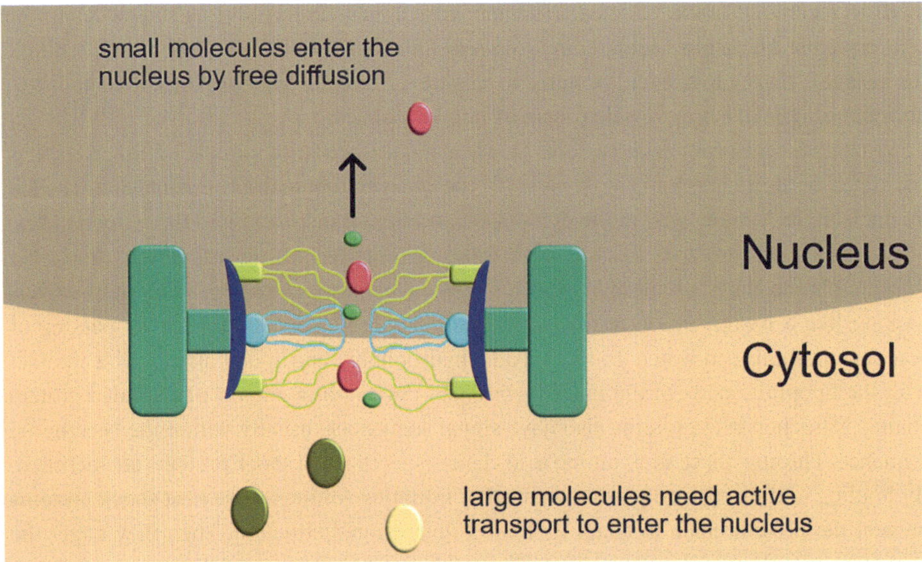

Fig. 3.16 Schematic representation of nuclear pore penetrating the nuclear double membrane (not shown): Large protein complexes outside of pore (dark cyan and blue), FG proteins fill inner pore with long unstructured FG-repeat chains (cyan and yellow threads)

With these NLS signals, proteins bind importins, which interact with FG-repeats and wiggle their way through the pores, delivering the protein cargo to the nucleoplasm (Fig. 3.17). Vice versa, export receptors bind to proteins in the nucleus via export signals and move them out of the nucleus. The necessary energy is provided by GTP with the help of small G-proteins of the ran-family (see Box 3.3 for explanation of G-protein function). GTP-bound ran-GTPases interact with importin-protein complexes. When GTP is hydrolyzed importin is released form the GTPase and attaches to the NLS of a cargo protein. This complex is transported through the nuclear pore. In the nucleoplasm, ran-GTP is present, which captures importin, thereby releasing the cargo protein into the nucleoplasm. ranGTP-importin may either be directly exported through the nuclear pore into the cytoplasm, where GTP is hydrolyzed and the cycle starts anew to transport another molecule into the nucleus (Fig. 3.17). Alternatively, ranGTP can interact with exportin molecules attaching to proteins with nuclear export signal (NES) and transport the resulting complexes out of the nucleus. After GTP hydrolysis, the complex breaks up and the protein cargo is released into the cytoplasm.

The activity of nuclear proteins can be regulated by controlling their access to the nucleus. For example, nuclear localization signals can be hidden inside a molecule and only be released after a change in protein structure which may come about by phosphorylation at an allosteric side. Similarly, nuclear localization signals can be blocked by protein–protein interactions with regulatory proteins and only released in response to signals. Moreover, some proteins are only transferred into the nucleus when they interact with an NLS-containing binding partner that gives them a piggyback.

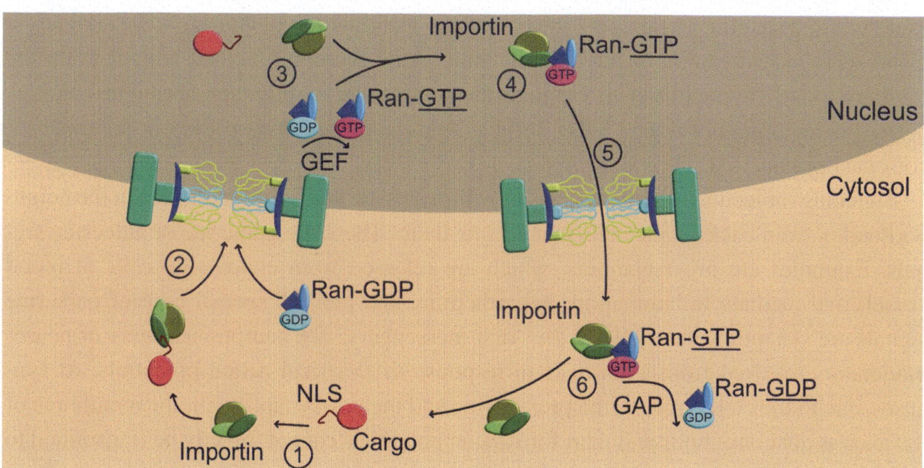

Fig. 3.17 Schematic illustration of nuclear import: Step 1: cargo protein (red) with NLS binds importin (green) protein complex in cytoplasm; Step 2: importin with cargo translocates into nucleus, RanGDP (blue/cyan symbol) diffuses into nucleus; Step 3: Inside nucleus, cargo is released; Ran-GEF binds Ran and exchanges GDP for GTP; Step 4: Ran-GTP binds cargo-less importin; Step 5: RanGTP binds importin and translocates back into cytoplasm; Step 6: RanGTP binds RanGAP in cytoplasm, GTP is hydrolyzed to RanGDP; RanGDP separates from importin and another cargo molecule can be loaded onto importin

3.7 Signal Transduction at Membranes: The Function of Seven-Transmembrane Domain Receptors

Signaling molecules can, when they are not hydrophobic, can not penetrate the plasma membrane to directly transmit their message. Instead, they interact with membrane receptors, which then "transduce" the information to the inside of the cell.

The major class of membrane receptors in animals are so-called seven-transmembrane domain receptors, (7 TMD receptors), (Fig. 3.3). These are integrated membrane proteins, which, as the name suggests, span the plasma membrane seven times. Their N-terminal end is directed toward the outside of the cell, and the C-terminal end extends into the cytoplasm. The transmembrane regions are dominated by hydrophobic amino acids and connected by hydrophilic protein loops. Three hydrophilic loops of the protein chain are directed outwardly of the cell and three are directed inwardly.

Inside of the cell, seven-transmembrane domain receptors engage a transduction system, which reacts to extracellular signals binding. The transducers are G-proteins, and therefore, seven-transmembrane receptors are also called G-protein-coupled receptors (GPCRs). These G-proteins often activate enzymes, which then produce small molecules that are distributed in the cell or at the inside of the plasma membrane and in turn activate appropriate cellular mechanisms in response to the external signals. Such small molecules are called "second messengers", as opposed to the actual external signaling molecule, which activates the receptor in the first place and therefore constitutes the "first messenger".

External signals in the human body can, for instance, be endocrine signals, which are secreted hormones traveling in the blood before they reach their target cells. These are excreted from specific glands, e.g., the pancreatic gland secreting glucagon, the adrenergic gland secreting adrenalin, or the hypothalamic gland secreting factors, which act on the pituitary gland (hypophysis) to regulate the production of pituitary hormones such as thyroid-stimulating hormone (TSH), follicle-stimulating hormone (FSH), or adrenocorticotropic hormone (ACTH).

Signaling molecules are often released within a tissue and they act on cells in the neighborhood or even back on the cells that secrete them. These are paracrine or autocrine signals. Examples are prostaglandins, which are released from endothelial cells of blood vessels and mediate inflammation, hyperthermia, and pain. A special case of paracrine signals are synaptic signals. Here, the "first messengers" are neurotransmitters or neuromodulators released from axon ends in response to electrical action potentials. At synapses, due to the cleft between the presynaptic and the postsynaptic cell, the membrane of the axon is not continuous and therefore action potentials cannot directly be transmitted to the dendrite of the next nerve cell. Therefore the electrical signal is transformed into a chemical signal, which travels the synaptic cleft and transmits the information to the dendrite. This occurs by neurotransmitters binding the dendritic membrane and inducing again action potentials on dendrites, spreading the signal into the next neuron. The receptors on the postsynaptic dendritic membranes are often ion channels, which are neurotransmitter gated, meaning they open in response to neurotransmitter binding. However, seven-transmembrane domain receptors are also used for neurotransmission. They lead to the production of second messengers, which open ion channels from the inside of the cell to permit the original action potential. Some seven-transmembrane domain receptors are located close to the synaptic cleft within neuronal membranes. When activated, they work as modulators of neurotransmission. Examples are the feedback inhibition of neurotransmission by the endocannabinoid system and the modulation of pain perception through descending pain pathways employing opiate receptors.

Furthermore, our entire vision depends on a seven-transmembrane domain receptor, which will be detailed later. Similarly, some taste and all olfactory sensing is mediated by seven-transmembrane domain receptors. As these examples show, there are many and diverse classes of molecules engaging seven-transmembrane receptors as external signals, including hormones (adrenalin), neurotransmitters (also adrenalin, acetylcholine, dopamine, serine), peptides (blood pressure regulators such as angiotensin and bradykinin, neuromodulators such as enkephalins and endorphins), and the many olfactory receptors securing our senses of smell and taste.

A specific class of G-proteins is dedicated to seven-transmembrane receptor signaling. These are trimeric, meaning that they consist of three subunits, α, β, and γ. The α-subunit is the actual G-protein (remember, this is a protein that is able to bind GTP and to hydrolyze GTP to GDP, it needs a Guanylnucleotide exchange factor GEF to loosen GDP and bind GTP, and it often needs help to hydrolyze this GTP to GDP, which is provided by a GTPase-activating protein GAP, see Box 3.3).

In the process of G-protein-coupled receptor (GPCR) signaling, the receptor after activation by a signaling molecule works as GEF, and so-called *r*egulators of *G*-protein *s*ignaling (*RGS* proteins) work as GAP. After the signaling molecule has bound, the seven-transmembrane receptor changes its conformation in such a way that it interacts with the α-subunit of the trimeric G-protein and removes GDP, whereupon GTP binds to it. GTP hydrolysis, in contrast to small G proteins, can be carried out by the relatively high intrinsic GTPase activity of the α-subunit of trimeric G proteins. Nevertheless, trimeric G proteins are also regulated by GAPs; here, they are called *RGS* proteins (regulators of G-protein signalling).

When GTP binds to the trimeric G-protein, the protein complex dissociates and releases the GTP-bound α-subunit. The β- and γ-subunits remain together. The GTP-bound and thus activated α-subunit is now able to interact with downstream targets, which then produce second messengers (Fig. 3.18). An important target for GTP-bound Gα-subunits is the adenylate cyclase, the enzyme that converts ATP to cAMP. Two members of the trimeric G-protein family regulate it, the stimulatory G-protein G_Sα-subunits activate it and the inhibitory G-protein G_iα-subunits inhibit it. Thus, either more cAMP is produced or less cAMP is produced in response to the respective receptor activation (Fig. 3.19). Target for the α-subunit of the trimeric G-protein G_q is the phospholipase C, which produces inositol-3-phosphate (IP3) and diacylglycerol (DAG), whereby DAG remains in the membrane and IP3 opens Ca^{2+}-channels in the ER (see Box 3.4). Finally, the α-subunit of the trimeric G-protein, transducin, activates a phosphodiesterase degrading cGMP to regulate the activation of photoreceptor cells in our visual system.

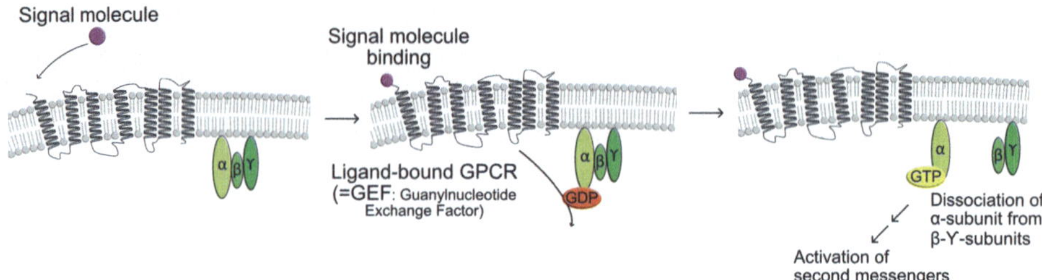

Fig. 3.18 G-protein coupled (seven-transmembrane domain)-receptor signaling; seven-transmembrane domain receptor with transmembrane helices, extracellular signaling molecule (purple), trimeric G-protein with α, β, and γ-subunits (shades of green) are shown: From left to right: inactive receptor, signaling molecule binds receptor, activated receptor obtains GEF activity for α-subunit of G-protein, GDP is removed, GTP binds, and Gα dissociates from βγ-subunits. Gα activates target proteins including enzymes, which then produce and activate second messengers

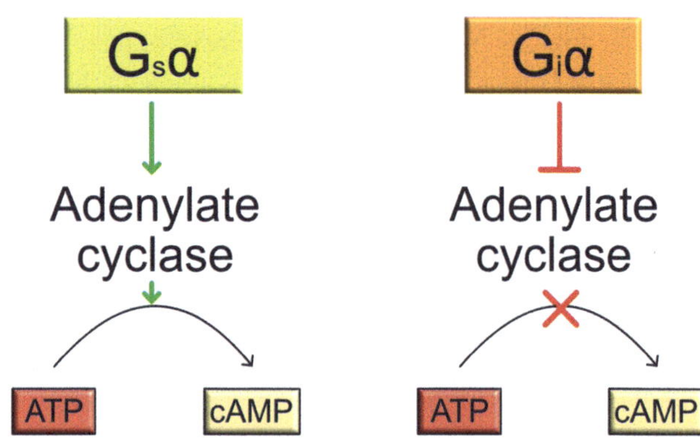

Fig. 3.19 Schematic illustration of stimulatory G-protein $G_S\alpha$-subunit (yellow box) inhibitory G-protein $G_i\alpha$-subunit (orange box) activating or inhibiting the conversion of ATP to cAMP by adenylate cyclase

Box 3.4 Ca^{2+}-Tool Kit

Ca^{2+} -ions play an important role in many biological processes. The concentration of Ca^{2+} -ions in the cytoplasm is, however, very low in comparison with the concentration outside the cell. The ER and mitochondria have a higher concentration of Ca^{2+} -ions than the cytoplasm, but a lower one than the extracellular space. In order to maintain the low cytoplasmic Ca^{2+} -ion concentration, cells use *ATP-driven Ca^{2+} -pumps* to remove Ca^{2+} -ions out of the cell and into the ER. These transporters demand a large proportion of the cell's energy. Moreover, Ca^{2+} -ion-binding proteins such as calmodulin in the cytoplasm and calreticulin in the ER, work as Ca^{2+} -ion buffers and ensure a low concentration of freely diffusible Ca^{2+} -ions. In addition, cells have Ca^{2+} -release channels, which release Ca^{2+} -from the extracellular space or from the ER into the cytoplasm in response to stimuli. The inositol-3-phosphate (IP3)-gated ER Ca^{2+} -channel is one example for a second messenger gated Ca^{2+} -channel. In addition, there are voltage-gated Ca^{2+} -channels and ligand operated Ca^{2+} -channels. Moreover, when Ca^{2+} -storage compartments, like the ER, get

Box 3.4 (continued)

low on Ca^{2+} -ions, storage-operated channels are able to refill them with extracellular Ca^{2+}. Many cellular processes are directly regulated by Ca^{2+} -ions. These include muscle contraction (Chap. 4), neurotransmission, secretion (this chapter), fertilization, biomineralization, and others. Many enzymes are regulated by Ca^{2+} -ions, for instance, protein kinases, such as protein kinase C, calmodulin-dependent kinases, myosin light-chain kinase, and proteases like calpain, and others, to mention just a few.

3.7.1 Adenylate Cyclase

The activity of adenylate cyclase regulates the concentration of cAMP in the cell. cAMP is a regulator of several cellular pathways. It regulates the activity of protein kinase A (PKA), which in turn may phosphorylate many metabolic and transcriptional regulators. It also regulates the opening of ion channels from the inside of the cell and it activates the GEF for small G-proteins of the Ras superfamily, which are called rap proteins (*ra*s-related *p*rotein) (Fig. 3.20). Like ras proteins, they function in cellular regulatory processes controlling cell division.

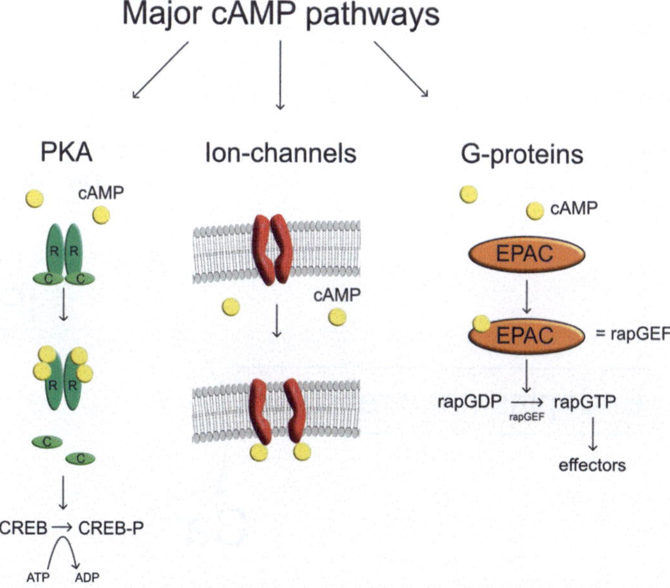

Fig. 3.20 Schematic illustration of the three major pathways activated by cAMP: Activation of cAMP-regulated protein kinase (Protein kinase A, PKA), opening of cyclic nucleotide-gated ion channels (CNC) and activation of exchange proteins directly activated by cAMP (EPACs, rapGEF is shown as an example for a cAMP-bound EPAC)

3.7.2 Phospholipase C

A certain class of G-proteins, named Gq, targets the enzyme Phospholipase C. Phospho*l*ipase *C* (*PLC*) modifies phospholipids, for instance phosphoinositols. Phosphoinositols are present in the lipid bilayer facing the cytoplasm and have important signalling functions. Phospholipase C hydrolyzes a phospho–ester bond of phosphoinositols, for example, PI 4,5-bisphosphate (PIP_2). By this activity of phospholipase C, inositol-3-phosphate (IP_3) is released from PIP_2 and a *dia*cyl*g*lycerol (DAG) remains in the membrane. Inositol-3-phosphate binds to Ca^{2+}-channels on the membrane of the ER allowing efflux of Ca^{2+} from the ER into the cytoplasm (Fig. 3.21). Ca^{2+} -ions can then

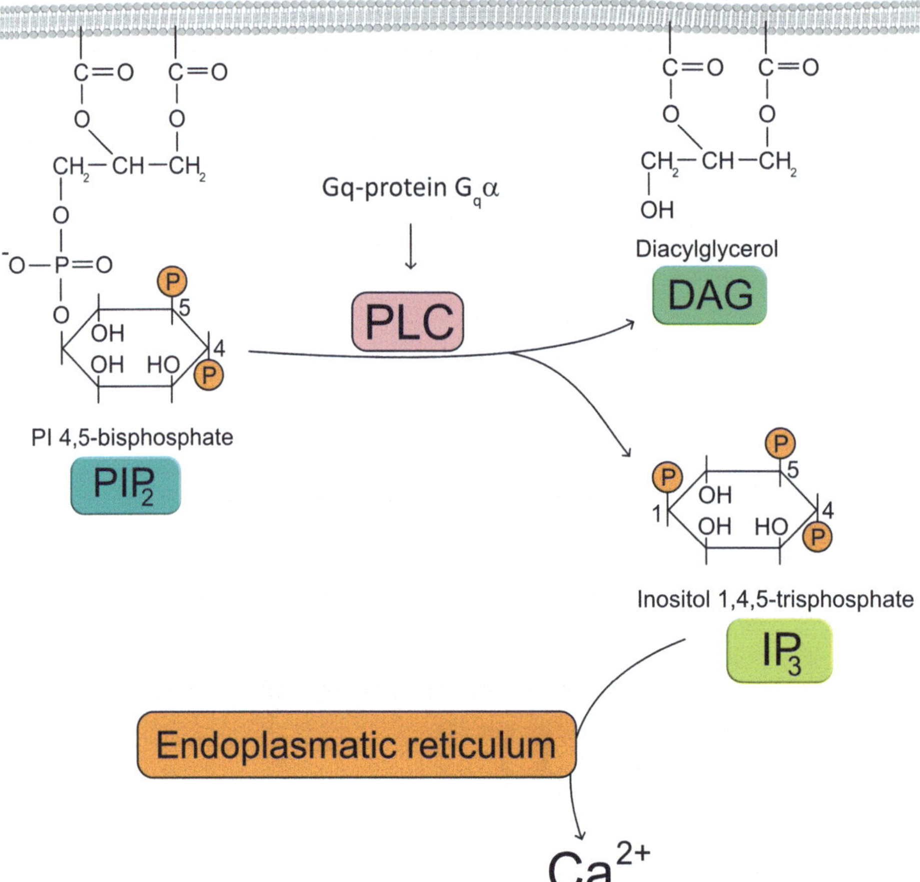

Fig. 3.21 Gq-protein $G_q\alpha$ activates phospholipase C. Phospholipase C (PLC, pink box) cleaves Phosphoinositol-4,5 biphosphate (PIP_2), blue box, to produce DAG, which remains connected to two fatty acids in the membrane (green box). Inositol-3-phosphate (IP_3, yellow box) is released into cytoplasm and binds IP_3-gated Ca^{2+} ion channel in the membrane of the ER, and this leads to release of Ca^{2+}-ions from the ER into the cytoplasm

function as second messengers. They are involved in regulating many enzyme activities. Together with diacylglycerol, for instance, they regulate the activity of protein kinase C, a kinase involved in growth control (see Box 3.4).

3.7.3 Photodetection of Retina

As an example for signal transduction at GPCRs, the process of photodetection in the retina is described. It has some very specific properties, such as low noise, high sensitivity (it may detect single photons), and a quantifiable outcome between 0.01 photons/μm^2/s in rod cells and up to 10^9 photons/μm^2/s in cone cells. It is also very fast. When light reaches our eye, it penetrates the lens and the image is displayed on the retina. The retina is packed with photoreceptor cells, which are connected to neurons. These neurons are bundled in the optic nerve, which connects with our brain. Photoreceptor cells are very long cells basally attached to the retinal epithelium. Their apical region contains a synaptic body connecting it with dendrites of nerve cells (Fig. 3.22). These chemical synapses use the

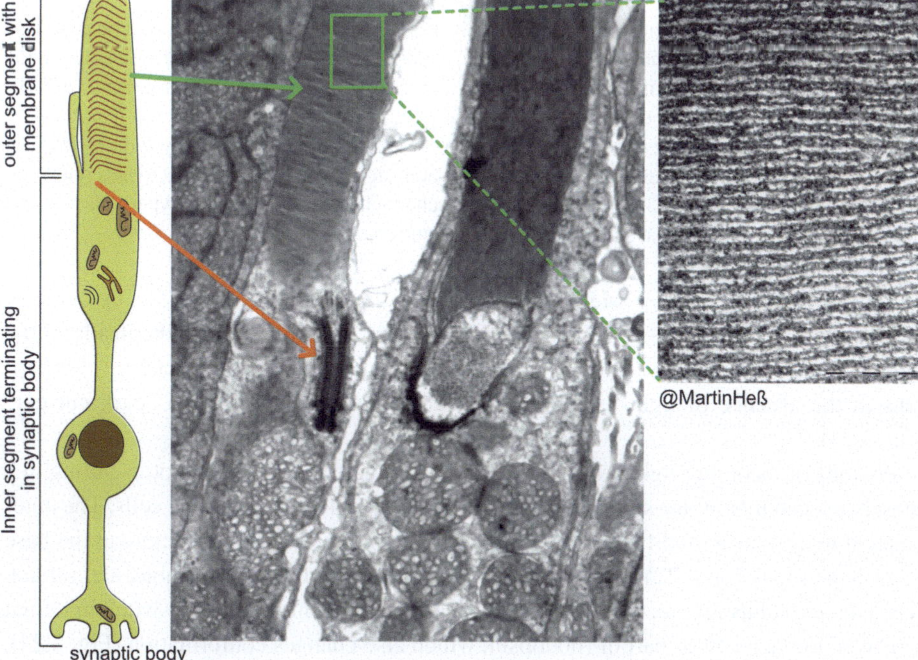

Fig. 3.22 Rod photoreceptor cell: TEM section of anchovy photoreceptor cell with inner and outer segments, inner segment contains nucleus, mitochondria, ER, and Golgi apparatus (brown symbols) and ends in synapse connecting to nerve cell, outer segment corresponds to a modified immobile cilium with microtubule base (orange arrow) and contains membrane discs (green arrow); insert shows magnification of membrane discs (scale bar 0.2 μm)

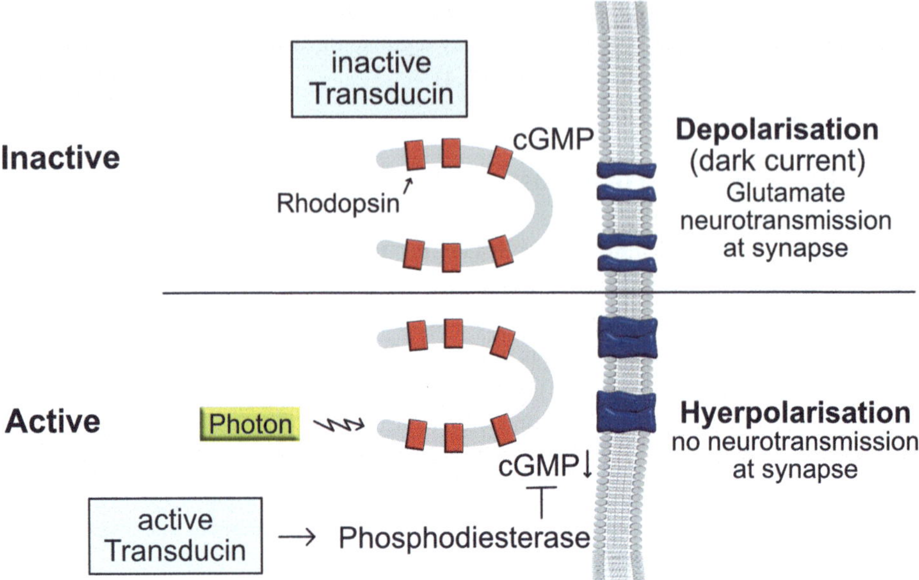

Fig. 3.23 Schematic illustration of rhodopsin activation: upper panel: Rhodopsin (red box) on disc membranes (gray) with one molecule on each disc, rhodopsin on upper membrane is inactive in the dark, G-protein (transducin) is inactive, phosphodiesterase is inactive, cGMP is present, cGMP-gated Na^+channels (blue) in outer cell membrane of photoreceptor cell are open, and membrane is depolarized (dark current); lower panel: Rhodopsin on membrane is activated by light (photon in yellow box), G protein (transducin) on membrane is activated (GTP bound) and activates phosphodiesterase, which degrades cGMP, cGMP-gated ion channels close, membrane hyperpolarizes and glutamate neurotransmission at synapse ceases (dark current stops)

neurotransmitter glutamate. Membrane depolarization is induced by the opening of cGMP-gated Na^+ and Ca^{2+}-ion channels. cGMP opens the channel from the inside of the cell. Na^+ and Ca^{2+}-ions flow into the cell, leading to membrane depolarization. This happens in the absence of light leading to constant glutamate signaling, a "dark current" (Fig. 3.23).

Photons of incoming light are sensed by 7 TM domain rhodopsin receptors sitting in a densely packed membrane system in the outer segment of the rod and cone cells. The outer segment represents a modified cilium with an immobile microtubule structure at the base (see Chap. 4 and Fig. 3.22). Rhodopsin receptors are bound to a small molecule, retinal. When a photon hits retinal, it changes conformation. The conformational switch in retinal is relayed to the receptor part of rhodopsin, which also changes conformation (Fig. 3.24). Now the activated receptor becomes a GEF for the trimeric G-protein transducin. The transducin α-subunit binds GTP and is released from its trimeric complex, leaving the βγ-dimer behind. GTP-bound transducin activates a cGMP-phosphodiesterase, which converts cGMP to GMP (Fig. 3.24). This decrease in cGMP leads to the closing of the cGMP-gated ion channels, and the "dark current" glutamate signaling to the brain is stopped. The photoreceptor cell is hyperpolarized. During this process, the original light signal derived

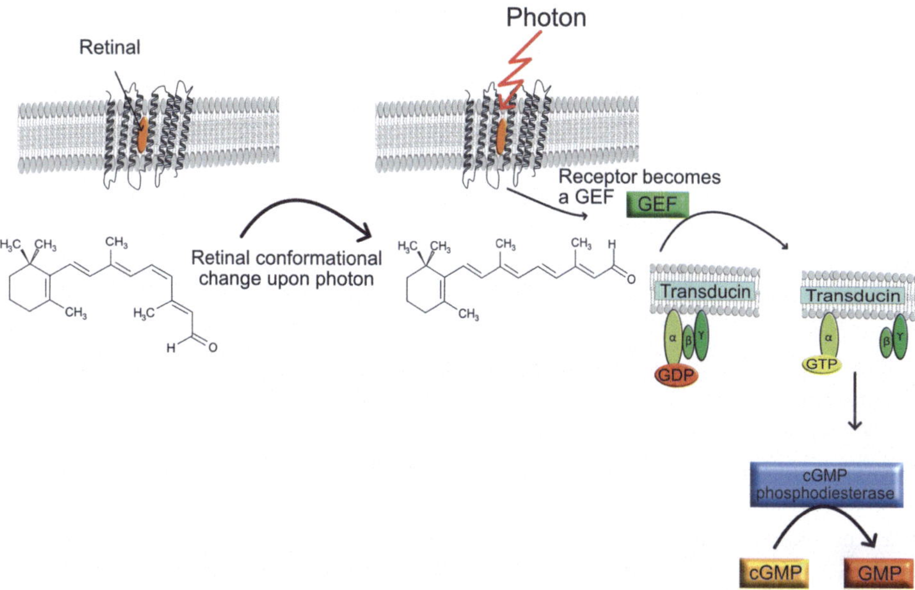

Fig. 3.24 Schematic illustration of rhodopsin signaling mechanism: Rhodopsin is bound to retinal, retinal changes conformation in response to activation by photon, receptor changes conformation and becomes GEF for the rhodopsin-specific trimeric G-protein transducin, transducin binds GTP, GTP-bound α-subunit dissociates and activates phosphodiesterase; cGMP is hydrolyzed to GMP

from one photon is immensely amplified. The photon hits the receptor turning it into the GEF for transducin. The activated receptor may exchange GTP for GDP on ca. 500 transducing molecules. Each transducing GTP activates one molecule of phosphodiesterase. However, one phosphodiesterase can hydrolyze ca. 2000 molecules of cGMP. These induce the closing of ion channels. To end the signal, there are two switches. First, the receptor is inactivated, which happens by binding of a "stopper" called arrestin, to its intracellular domains. Arrestin binding is preceded and conditioned by phosphorylation of the receptor by protein kinase A or other kinases. Second, GTP is hydrolyzed by transducin, which may be additionally activated by GTPase-activating proteins, called regulators of G-protein signaling (RGS). The more potent they are, the faster does the signal end. They are much more potent in cone cells than in rod cells; therefore, as we know, a light that hits our eye in the dark lingers much longer than light in daylight where our optical impressions change very fast. Also, night vision by rod cells is far more sensitive than day vision. In the dark, we can even see faint stars. This is possible because the signal amplification in the dark is higher, mainly caused by a slower inactivation of the GEF activity of the receptors, so GTP transducin can activate many more molecules of phosphodiesterase before it falls off the receptor.

References for original work and additional information concerning the above-described processes can be found in Chapter 3 of Böttger et al. (2018).

3.7.4 Toxins, Drugs, and Pharmaceuticals Acting on GPCRs

GPCRs are the most abundant targets for the development of pharmaceutical drugs. Many plant or bacterial products interfere with GPCRs as well. In the following, only a few commonly known examples are named.

3.7.4.1 Cholera and Pertussis Toxins

Two well-known bacterial toxins target G-proteins. These are cholera toxin and pertussis toxin, both causing severe disease. Cholera toxin is an ADP- ribosyltransferase produced by the bacterium *Vibrio cholerae*. It transfers one ADP-ribosyl onto an arginine residue within the active site of the GTPase, and this inhibits the GTPase activity of the $G_s\alpha$-subunit of a G_S-trimeric G-protein within the epithelium of the intestine. Therefore, cAMP concentration increases, leading to diarrhea. Pertussis toxin is an ADP ribosyltransferase produced by *Bordella pertussis* and targets an inhibitory G-protein, G_i. ADP-ribosyl is transferred onto a terminal cysteine residue of G_i blocking its interaction with the receptor. Now G_i is inhibited and the adenylylcyclase is free to increase the concentration of cAMP in the epithelium of the lung, causing the symptomatic whopping cough.

3.7.4.2 Atropin and Muscarine, Opiates and Morphin, and Cocain and Caffeine

Some important natural compounds interfere with acetylcholine receptors, and this concerns the GPCR-type acetylcholine receptors, which are also called muscarinic receptors because they are activated by muscarine (in contrast to nicotinic acetylcholine receptors, which are ion channels and are activated by nicotine). Muscarinic acetylcholine receptors are present in our vegetative nervous system where acetylcholine is released from preganglionic and postganglionic neurons of the parasympathetic nerve. Muscarin is a potent parasympathomimetic, causing narrowing of pupils, saliva flow, intestinal contractions, sweating, and circulation collapse. It can also cause heart paralysis. Intoxication with muscarine can be reversed by atropin, an alkaloid from *Atropa Belladonna*, which inhibits acetylcholine receptor signaling. It is well known for its widening of pupils which can be used by ophtamologists for examination of the eyes. Atropin used to be a beauty medicine because of the widening of the pupils, which was considered attractive in women (Fig. 3.25).

Opiates are alkaloids of the morphine type obtained from the poppy plant *Papaver sativum*. These plant products are not the endogenous ligands for human opiate receptors. The endogenous ligands are peptides like the pentapeptides Leu- and Met-enkephalin and endorphins. They activate opiate receptors in the brain and are the key players in our physiological pain control system. When pain stimuli, such as a cut in the finger, activate

Fig. 3.25 Photograph of *Atropa belladonna*. (**a**) Flower and (**b**) Immature fruit, which will ripen into blue berry

ascending pain pathways through a 3-neuron system using glutamate neurotransmission at synapses, simultaneously, descending pathways are activated, which are transmitted by the neurotransmitters serotonin and noradrenalin and end in spinal interneurons. These release enkephalins as a response, which dampens pain transmission. Enkephalins signal (among others) through an inhibitory G-protein Gi, causing a decrease in cAMP in the cells. Morphin and other opiate like plant-derived drugs belong to the most powerful pain killers we know; however, on the back side, they cause addiction and tolerance (meaning you need more and more of the drug to obtain the same effect). Moreover, their withdrawal is accompanied by severe withdrawal symptoms including hyperanalgesia, intestinal contractions, and others. It is speculated that this is due to hypersensitivity of adenylate cyclase induced by repeated stimulation of opiate receptors with morphine alkaloids.

Cocaine is present in the leaves of the coca plant, which is found in South America. It does not directly bind to a seven-transmembrane receptor; however, it blocks the transporter for dopamine, serotonin, and norepinephrine. These neurotransmitters are therefore not transported back into the presynaptic cell after their release but remain in the synaptic cleft, and with regard to dopamine, cocaine induces a "dopamine overflow." This provokes feelings of euphoria in the central nervous system. People living at high altitude in the Andes chew cocaine leaves, which protect from mountain sickness and stimulate the respiratory center. Interestingly, cocaine was the first local anesthetic used for dental treatment. Chewing cocaine apparently makes the mouth feel numb and the American surgeon William Stewart Halsted injected it near the nerves of the lower jaw before applying painful treatments to his dental patients. Before doing so, he had conducted cocaine experiments on himself and became severely addicted to it (Wikipedia, 2023). Cocaine later provided a chemical lead compound for the development of synthetic local anesthetics including procaine (Novocain) and lidocaine.

In humans, adenosine receptors are suggested to be the major cellular target for caffeine. Adenosine receptors are inactivated by very low concentrations of caffeine (1–10 μM). They respond to extracellular adenosine in nerve and glia cells, which is directly connected to the energy demand of those cells. When the intracellular ATP concentration lowers, adenosine concentration increases and it is transported out of the cell to activate the receptors. Those activated receptors induce an appropriate body response including decrease in motor activity and induction of sleep. By overriding these mechanisms, caffeine can keep us awake.

Original references for the drugs mentioned above and additional information can be found in Chapter 5 of Böttger et al. (2018).

Test Yourself

1. What are the components of a biological membrane?
2. Can proteins move within the membrane?
3. Which molecules can cross a biological membrane by passive diffusion?
4. Why are ion channels, transporters, and pumps needed?
5. Explain the membrane potential. How is it produced and maintained?
6. What is the difference between chemical and electrical synapses?
7. How do secreted or membrane proteins get into the ER?
8. What is the function of a "signal recognition particle"?
9. How are proteins transported from ER to Golgi apparatus? Does it also work vice versa?
10. What are the components that mediate vesicle budding from donor membranes, transport, and fusion with target membranes?
11. What are the principles of protein transport into membranes and in and out of the nucleus?
12. Explain the different activities of cholera- and pertussis toxin. Why do both activate adenylate cyclase?
13. Which receptors do muscarine, atropine, morphin, and caffeine bind to?
14. How does cocaine affect the mood?

References

Balch, W. E., & Rothman, J. E. (1985). Characterization of protein transport between successive compartments of the Golgi apparatus: Asymmetric properties of donor and acceptor activities in a cell-free system. *Archives of Biochemistry and Biophysics, 240*(1), 413–425.

Blobel, G. (2000). Protein targeting (Nobel lecture). *ChemBioChem, 1*(2), 86–102.

Böttger, A., Vothknecht, U., Bolle, C., & Wolf, A. (2018). *Lessons on Caffeine, Cannabis & Co: Plant derived drugs and their interaction with human receptors*. Springer.

Chua, N. H., Blobel, G., Siekevitz, P., & Palade, G. E. (1976). Periodic variations in the ratio of free to thylakoid-bound chloroplast ribosomes during the cell cycle of Chlamydomonas reinhardtii. *The Journal of Cell Biology, 71*(2), 497–514.

Dowhan, W. (1997). Molecular basis for membrane phospholipid diversity: Why are there so many lipids? *Annual Review of Biochemistry, 66*, 199–232.

Engelman, D. M. (2005). Membranes are more mosaic than fluid. *Nature, 438*(7068), 578–580.

Fujimoto, T., & Parmryd, I. (2016). Interleaflet coupling, pinning, and leaflet asymmetry-major players in plasma membrane nanodomain formation. *Frontiers in Cell and Development Biology, 4*, 155.

Gilmore, R., Walter, P., & Blobel, G. (1982). Protein translocation across the endoplasmic reticulum. II. Isolation and characterization of the signal recognition particle receptor. *The Journal of Cell Biology, 95*(2 Pt 1), 470–477.

Gorter, E., & Grendel, F. (1925). On bimolecular layers of lipoids on the chromocytes of the blood. *The Journal of Experimental Medicine, 41*(4), 439–443.

Kalderon, D., Roberts, B. L., Richardson, W. D., & Smith, A. E. (1984). A short amino acid sequence able to specify nuclear location. *Cell, 39*(3 Pt 2), 499–509.

Levental, I., Levental, K. R., & Heberle, F. A. (2020). Lipid rafts: Controversies resolved, mysteries remain. *Trends in Cell Biology, 30*(5), 341–353.

Lim, R. Y., Aebi, U., & Fahrenkrog, B. (2008). Towards reconciling structure and function in the nuclear pore complex. *Histochemistry and Cell Biology, 129*(2), 105–116.

Rothman, J. E. (2002). Lasker Basic Medical Research Award. The machinery and principles of vesicle transport in the cell. *Nature Medicine, 8*(10), 1059–1062.

Schekman, R. (2002). Lasker Basic Medical Research Award. SEC mutants and the secretory apparatus. *Nature Medicine, 8*(10), 1055–1058.

Schekman, R. (2010). Charting the secretory pathway in a simple eukaryote. *Molecular Biology of the Cell, 21*(22), 3781–3784.

Scott, D. C., & Schekman, R. (2008). Role of Sec61p in the ER-associated degradation of short-lived transmembrane proteins. *The Journal of Cell Biology, 181*(7), 1095–1105.

Simons, K., & Ikonen, E. (1997). Functional rafts in cell membranes. *Nature, 387*(6633), 569–572.

Singer, S. J., & Nicolson, G. L. (1972). The fluid mosaic model of the structure of cell membranes. *Science, 175*(4023), 720–731.

Skou, J. C., & Esmann, M. (1992). The Na,K-ATPase. *Journal of Bioenergetics and Biomembranes, 24*(3), 249–261.

Südhof, T. C., De Camilli, P., Niemann, H., & Jahn, R. (1993). Membrane fusion machinery: Insights from synaptic proteins. *Cell, 75*(1), 1–4.

Ungewickell, E., & Branton, D. (1981). Assembly units of clathrin coats. *Nature, 289*(5796), 420–422.

Walter, P., & Blobel, G. (1981a). Translocation of proteins across the endoplasmic reticulum. III. Signal recognition protein (SRP) causes signal sequence-dependent and site-specific arrest of chain elongation that is released by microsomal membranes. *The Journal of Cell Biology, 91*(2 Pt 1), 557–561.

Walter, P., & Blobel, G. (1981b). Translocation of proteins across the endoplasmic reticulum. II. Signal recognition protein (SRP) mediates the selective binding to microsomal membranes of in-vitro-assembled polysomes synthesizing secretory protein. *The Journal of Cell Biology, 91*(2 Pt 1), 551–556.

Walter, P., Ibrahimi, I., & Blobel, G. (1981). Translocation of proteins across the endoplasmic reticulum. I. Signal recognition protein (SRP) binds to in-vitro-assembled polysomes synthesizing secretory protein. *The Journal of Cell Biology, 91*(2 Pt 1), 545–550.

Wikipedia (2023). https://de.wikipedia.org/wiki/William_Stewart_Halsted

Shape and Structure of Cells and Tissues

4

What You Will Learn in This Chapter

Animal cells have a cytoskeleton which, on the one hand, provides stability, structure, form, and polarity. On the other hand, this cytoskeleton is very dynamic and controls cell and organelle movements, shape changes, and even cell division. The cytoskeleton in vertebrate cells consists of three components, comprising microfilaments with a diameter of ca. 7–9 nm, macrofilaments with a dimeter of 25 nm, and intermediate filaments with a diameter of ca. 10 nm, and the latter represents a size in between that of micro- and macrofilaments. All components of the cytoskeleton form networks of fibers inside the cytoplasm. All filaments are formed and degraded by assembly and disassembly of subunits. In case of macro- and microfilaments, these subunits are globular proteins. They have a molecular mass of ca. 50 kDa. In contrast, intermediate filaments are constructed of helical proteins that bundle up into strong fibers (Fig. 4.1). Moreover, in multicellular animals, cells are embedded in tissues and organs. In those structures, cells are connected with each other by cellular junctions, junctions also exist between cells and the extracellular matrix. We discuss tight junctions, adherens junctions, gap junctions, desmosomes and hemidesmosomes. In addition we look at the principal composition of the extracellular matrix with the components collagen, hyaluronic acid, proteoglycans, and anchoring proteins, and discuss the structure and biomineralization of dentin, a major component of our teeth.

A. Böttger, J. Moneer, *Basic Cell Biology for Dentists and Medical Students*,
https://doi.org/10.1007/978-3-032-03636-0_4

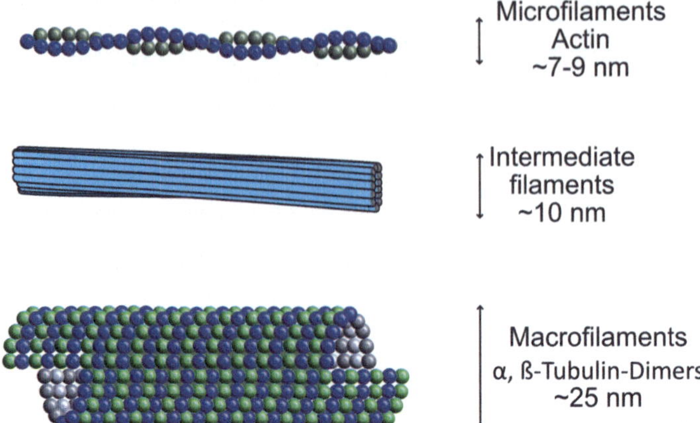

Fig. 4.1 Schematic representation of the three components of the cytoskeleton in vertebrate cells: Upper panel shows microfilaments consisting of two chains of globular actin molecules (actin indicated by blue and green spheres forming each chain) with a diameter of 7–9 nm; Middle panel shows intermediate filaments representing bundles of proteins with long α-helical secondary structure (indicated by thin light blue bars) with a diameter of 10 nm; Lower panel shows macrofilaments consisting of tubes with a diameter of 25 nm, formed by chains of α- and β-tubulin dimers (indicated by blue α-tubulin and green β-tubulin spheres)

4.1 The Cytoskeleton

4.1.1 Actin Microfilaments

Microfilaments are actin polymers, built from monomers of globular actin. Actin is highly abundant in all cells of the body and it is conserved throughout nature; it is also found in prokaryotes including bacteria and *Aarchaea* and even in bacteriophages (Pollard & Goldman, 2018). Microfilaments have many different functions: In epithelia lining the surface of kidneys and the gut, they form microvilli. These apical epithelia cell protrusions extend the surface of such epithelia to increase compound exchange (Fig. 4.2). In amoeboid cells, actin filaments constitute the main component of filopodia and lamellopodia, which are responsible for the amoeboid changes of cell shapes and movements. Moreover, actin fibers form a so-called adherence belt, which connects neighboring epithelial cells via adherens junctions. There are many other functions of actin filaments including the formation of stress fibers in response to environmental or cell intrinsic damage. By teaming up with motor proteins of the myosin family, actin fibers are involved in muscle contraction. They are also essential for cell division by forming, together with myosin, the contractile ring for cell constriction and separation of daughter cells (Fig. 4.2).

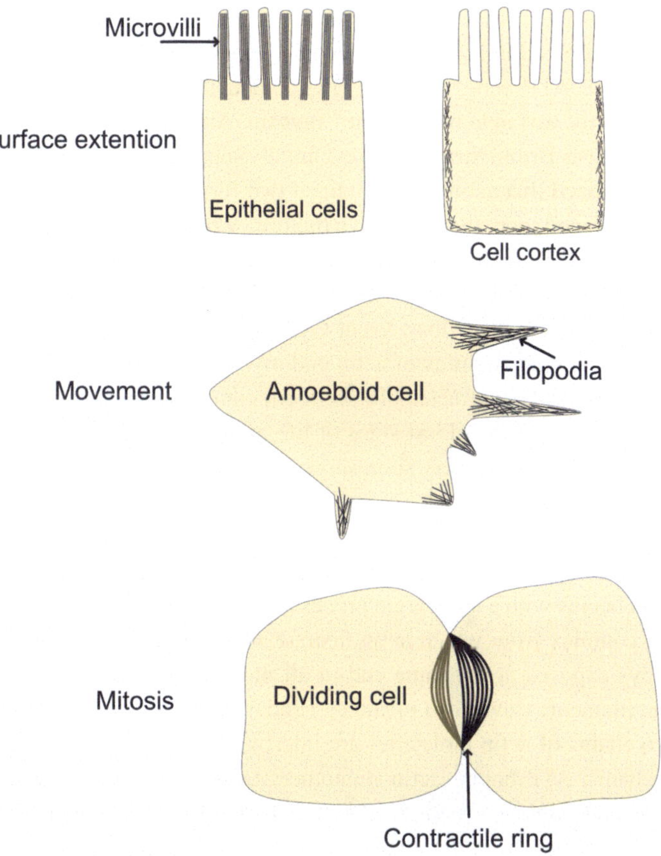

Fig. 4.2 Schematic representation of cells with microvilli formed by actin cables and actin cortex (upper panel); Amoeboid cells moving by forming filopodia (middle panel); Dividing cell in cytokinesis with contractile ring (lower panel); actin indicated by black lines

Actin was originally discovered in muscle fibers by the lab of Albert Scent-Györgyi in the thirties of the last century in Szeged, Hungary. In a review article, he reflects about the beginning of these discoveries: "There is one thing one can always do, and this I did: repeat the work of old masters" (Szent-Györgyi, 1963). The old master in this case was Albert W. Kühne who had reported about his isolation of the contractile proteins from frog muscle cells in 1864. He called the isolated substance "myosin" and characterized it chemically as protein, and physically as insoluble in alcohol, ether, or water, but dissolvable in neutral salt solutions. He found that the liquid extract of muscle fibers when filtrated through linen tissue obtained a syrup-like consistency, and coagulated, when dropwise released into water (Kühne, 1864).

Albert Scent-Györgyi and his co-worker Ilona Banka discovered, when extracting myosin from muscle cells with a salt solution that, depending on the time span for which the extraction was performed (20 min versus 24 h), two forms with substantially different viscosity were obtained. They referred to them originally as myosin A and myosin B. They also discovered that by adding ATP, only myosin B lost its viscosity, but could be "gelatinized" again, when ATP was removed. They called the more gelatin-like form actin, because it apparently was able to "activate" myosin. Another member of the same lab in Szeged, Hungary, was Bruno Straub, who eventually separated actin and myosin biochemically. He also realized that actin itself obtained two forms, one globular structure, which he called G-actin, and one filamentous structure, which he called F-actin (Bugyi & Kellermayer, 2020). Eventually, they were able to obtain thread-like structures from myosin, which started to contract, when actin was added. It is worth to reproduce here the account of this experiment that Albert Szent-Györgyi left us himself: "So we made threads of the highly viscous new complex of actin and myosin, "actomyosin," and added boiled muscle juice. The threads contracted. To see them contract for the first time, and to have reproduced in vitro one of the oldest signs of life, motion, was perhaps the most thrilling moment of my life. A little cookery soon showed that what made it contract was ATP and ions" (Szent-Györgyi, 1963).

Globular actin (G-actin), the monomer of actin filaments measures 5.5 × 5.5 × 3.5 nm. In the presence of Mg^{2+}- and K^{+}- or Na^{+}-ions, actin monomers polymerize into a chain. Polymerization begins with a nucleation process, where initially two or three G-actin molecules form a complex from which actin fibers evolve. Actin has a binding site for ATP, which is always exposed at the same end in all monomers of the chain, thus providing polarity to the filaments (given as (+) and (−) ends). In a single actin filament (or microfilament), two chains of actin molecules are intertwined, both with the same polarity. To prolong the chains, ATP-bound actin monomers (also called ATP-G-actin) are always added at the (+) end. At the (−) end, ADP-bound actin falls off the chain. The length of the filament therefore depends on the equilibrium between adding actin-ATP at the (+) end and releasing actin-ADP at the (−) end. Actin-ADP can also be added again to the (−) end, but this process of (−) end growth is usually slower than (+) end growth. By the adding of new G-actin at the (+) end of filaments, a flow-through of actin monomers within the polymer chain is established. This is called "treadmilling." Actin has an enzymatic ATPase activity. When a new actin-ATP monomer is added at the (+) end, the previously outermost actin-ATP moves one position down toward the (−) end of the filament and by doing so, it hydrolyzes ATP. However, the anorganic phosphate (P_i) is not released, but kept closely to the ATP-binding site (Fig. 4.3). Thus, the energy derived from ATP hydrolysis is stored in the filament and can be used at any time necessary, for instance, to quickly make filopodia for cells to move.

Obviously, the growth and decline of actin filaments are regulated by many factors. Here, only a few important ones are mentioned:

1. The Adenosine Nucleotide Exchange Factor "Profilin"
   ADP-actin has to be returned to ATP-actin so that it can be added again at the (+) end of growing filaments when needed. Very much as explained for GDP-binding to G-proteins (see Box 3.3 in Chap. 3), ADP binds very tightly to actin (otherwise, it would be easily replaced by ATP, which we have in large excess over ADP in a living cell).

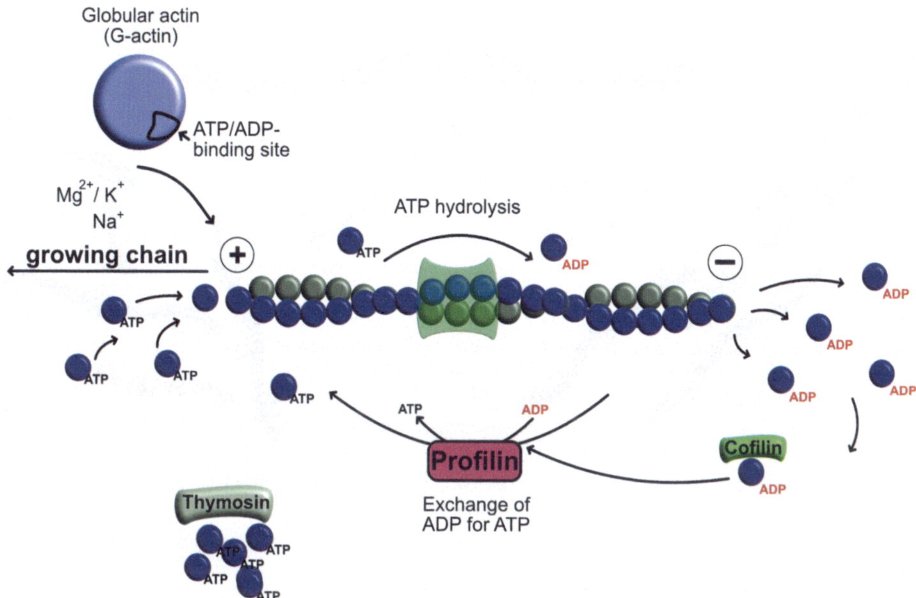

Fig. 4.3 Schematic representation of actin filaments composed of globular actin (blue and green spheres) with ATP/ADP binding site, actin filaments are built on the (+) side by addition of ATP actin and degraded on the (−) side whereby ADP actin is released (shown for blue actin polymer), factors regulating actin filaments include cofilin-binding ADP actin and transferring it to profilin, the adenosin nucleotide exchange factor profilin removing ADP from actin for ATP to bind, and ATP actin can be added to the (+) side of actin filaments or stored as a reserve by binding to the small protein thymosin

Therefore, an adenosin-nucleotide exchange factor is required, which removes ADP from its binding site on actin, thus allowing ATP to bind to the actin molecule. This is a function of profilin (Fig. 4.3). By binding ATP-actin, profilin also prevents spontaneous nucleation of new actin fibers, and adding of ADP-actin to the (−) ends. Instead, it delivers ATP-actin to (+) ends, thereby supporting filament growth into the (+) direction.

2. The ADP-Actin Binding Factor "Cofilin"
   Cofilin binds ADP-actin (Fig. 4.3). In this way, it prevents reassociation of ADP-actin with the (−) end and thus supports filament depolymerization.
3. The G-Actin Storage Factor "Thymosin"
   Thymosin also binds actin monomers to build a reservoir. It binds much stronger to ATP-actin than to ADP-actin (Fig. 4.3) and thus rather supports (+) end growth when required.
4. The Capping Proteins "CapZ" and "Tropomodulin"
   For most functions, actin filaments have to be restricted at a certain size. This means that no further ATP-actin should be added and/or no ADP-actin should be released. This is ensured by capping proteins like CapZ at (+) ends of actin filaments and tropomodulin at (−) ends of actin filaments. These proteins bind to the ends of the chains and

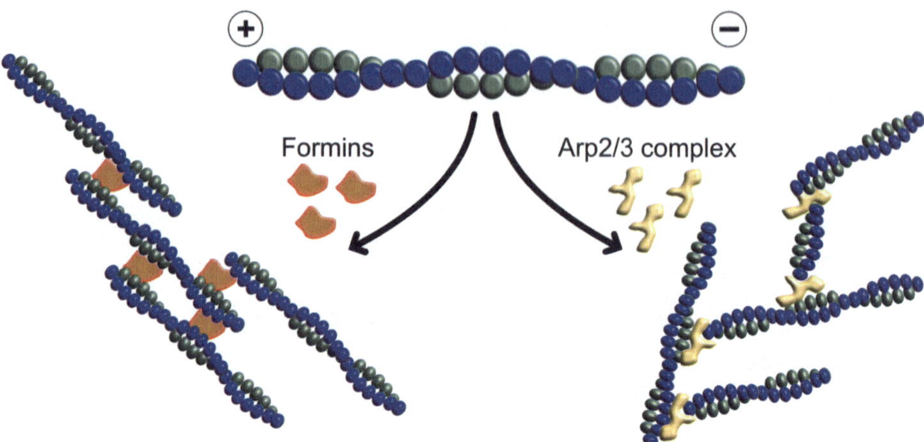

Fig. 4.4 Schematic representation of formins (orange shape) and Arp2/3 (yellow shape), formin connects multiple parallel actin chains to form bundles, Arp2/3 seeds new filaments from old ones to allow branch formation, actin chains are shown by blue and olive green spheres, and (+) and (−) ends are indicated

prevent actin monomers from entering or leaving the chains. The structure of the actin network in a cell depends on the activities of capping proteins because they determine the time when the filaments are free to grow.

5. The Branching Factors of the "Formin" Family and "Arp2/3"
   Arp2/3 is a protein complex that is associated with uncapped filaments at the side and nucleates the formation of branches of filaments, which are then blocked by capping proteins. This happens, for instance, during the formation of filopodia and lamellopodia during cell movement. Formins constitute a large protein family and block the + end of actin filaments from capping to enable the formation of very long filaments that can then be crosslinked into strong bundles and even cables. Arp2/3 and formins thus organize actin filaments into a network, which is important for most actin functions (Fig. 4.4).

Box 4.1 Toxins Targeting the Actin Cytoskeleton

Some toxins have an effect on actin, e.g., *phalloidin*, which is found in a poisonous mushroom, the death cap. It intercalates between actin subunits and thus prevents depolymerization of filaments. Fluorescently labeled phalloidine is often used by researchers to depict actin fibers in cells. The death cap *(Amanita phalloides)*, however, is deadly because it contains α-Amanitin in addition to phalloidine. Phalloidine is essentially destroyed when passing the stomach. α-Amanitin, in contrast, is more stable and inhibits transcription by RNAPol2, causing liver failure.

Latrunculin prevents formation of actin fibers, and it is found in some sponges. Cytochalasin D is a plant pathogen binding to (+) ends of actin fibers, thus preventing filament growth.

4.1.2 Muscle Actin and Muscle Contraction

Movement of skeletal muscle is carried out by myosin motor proteins shifting actin chains. Muscle fibers are the smallest unit of this machinery. In skeletal muscle, each muscle fiber is a large, elongated cell, which has several nuclei. This results from fusion of muscle progenitor cells during muscle cell differentiation into muscle fibers. Differentiated muscle cells are packed with myofibrils, which are fibers of actin arranged with myosin. Several such cells form a bundle and several bundles unite to form a muscle (Fig. 4.5).

Actin- and myosin - filaments in a muscle fiber are arranged in sarcomeres. Muscle cells thus contain a chain of sarcomeres, which can be recognized by electron microscopy as stripes. Hence, skeletal muscle is also called "striped muscle" (or striated muscle), in contrast to the "smooth muscle" present in the walls of visceral organs like intestine, pancreas and liver. Smooth muscle cells only carry one nucleus and have a spindle like shape. Moreover, actin-myosin filaments in smooth muscle cells are not arranged in sarcomers. Therefore these muscles do not appear striped. Heart muscle cells, in contrast, are striped, they have a special branched architecture. The branches of individual cells are interconnected by intercalating discs. Each heart muscle cell has one or two nuclei in its central part. In the following only skeletal muscles are considered.

The stripes of skeletal muscle represent stripes of actin-myosin fibers in a parallel order that are sideways restricted by strong disks, so-called z-disks. Actin-myosin filaments are firmly attached to these disks on both sides. When they contract, the distance between two z-disks is decreased. When they relax, it is increased. As shown in Fig. 4.6, myosin and actin filaments are alternately arranged. For muscle contraction, actin filaments are shifted in and out of stationary myosin filaments (Fig. 4.6).

Myosin 2, the myosin motor of skeletal muscle is composed of several subunits. The two larger of these subunits (heavy chains) have a long helical part each and together these helices form a coiled-coil structure like a rod. Each helix ends in a globular head region with an ATP-binding site. Smaller myosin subunits (light chains) are associated with the coiled-coil part directly behind the head, creating a neck. These have regulatory functions (Fig. 4.7).

For movement of muscle fibers, myosin heads attach to actin filaments and perform a stroke, which moves the filament towards the middle of the sarcomere. In this way, the

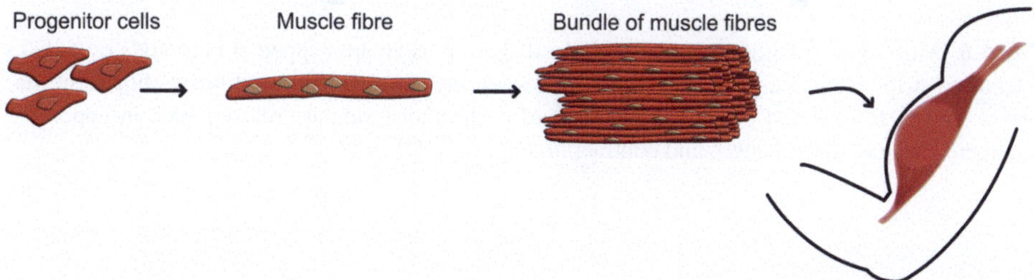

Fig. 4.5 Skeletal muscle structure: Each individual muscle fiber is formed by fusion of progenitor cells and contains multiple nuclei, muscle fibers arrange in bundles, muscles are structured by muscle fiber bundles

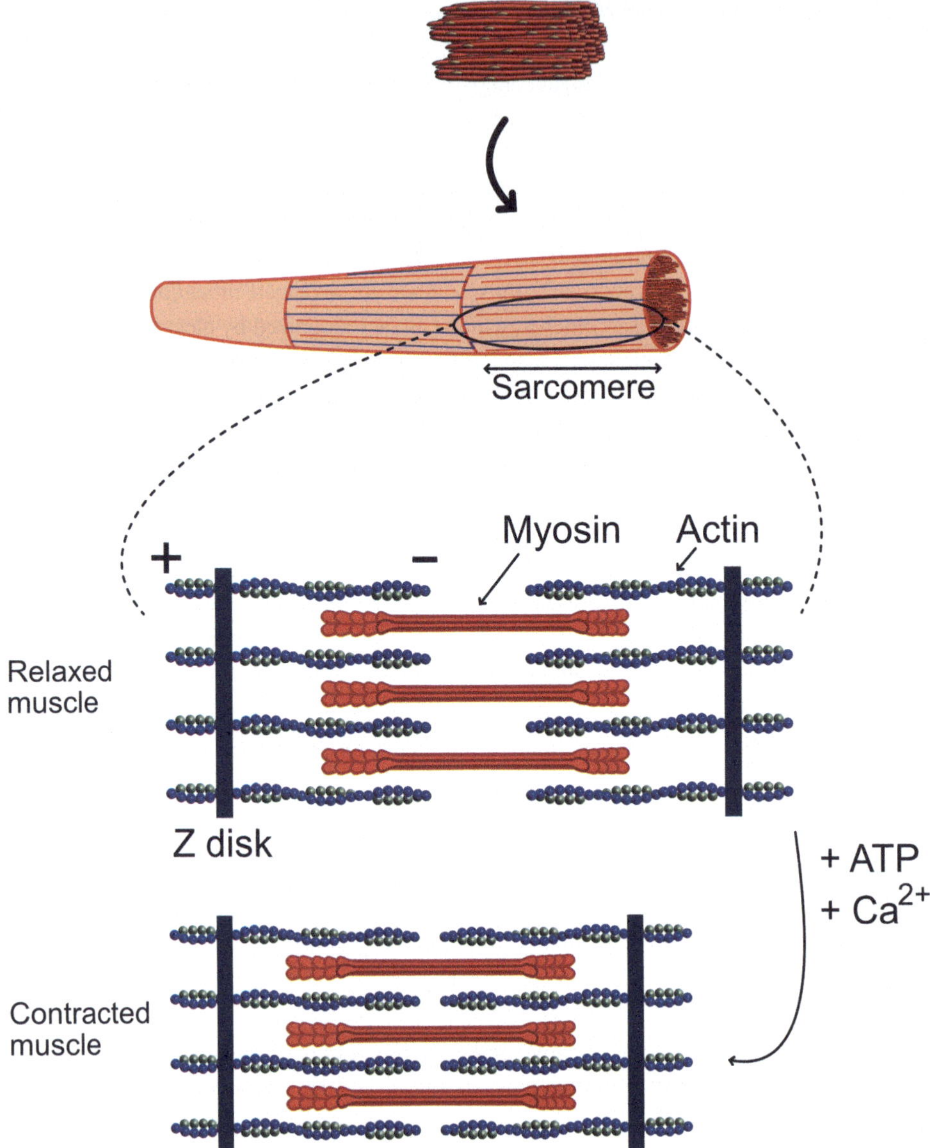

Fig. 4.6 Muscle contraction: Muscle fibers in skeletal muscle are capped at both ends by z-disks leading to striped appearance, muscle contracts when myosin filaments (red lines) shift actin filaments (chains of blue and green spheres) toward each other (compare relaxed state in upper and contracted state in lower panel), and contraction requires ATP and Ca^{2+}

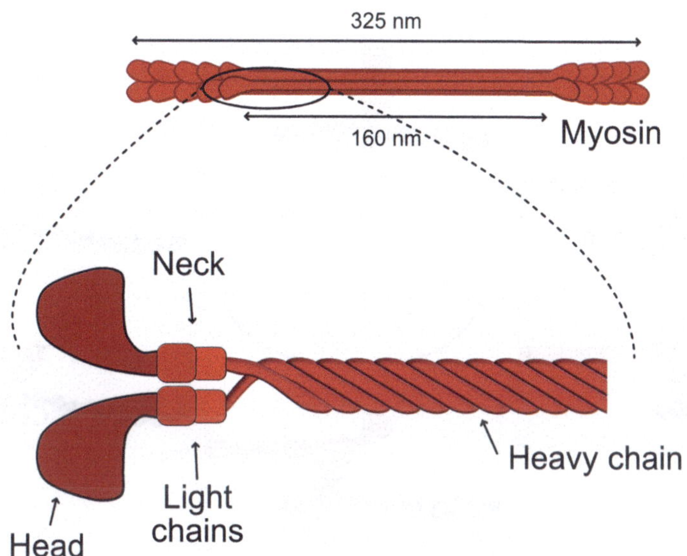

Fig. 4.7 Schematic representation of myosin (upper panel); Enlarged representation in lower panel shows heavy chains of myosin with globular head and helical tail structure and light chains attached to form neck structure

muscle shortens, or contracts. After each stroke, myosin heads bind ATP and detach from the actin filament. After detachment, this ATP becomes hydrolyzed, but P_i is not released, resulting in a high-energy state of the myosin head. In this state, the myosin head contacts the actin fiber. Then it releases P_i and the stored energy is released, resulting in a "*power stroke*" by which the myosin molecule stretches and moves the actin fiber (Fig. 4.8). Immediately after the power stroke, ATP is bound and the head detaches from actin (Cooke, 2004).

Skeletal muscles move in response to signals from the nervous system. Motor neurons are connected to skeletal muscle cells via a synapse - the neuromuscular junction. In the absence of a neuronal signal for muscle contraction, the actin binding site on myosin fibers is blocked by two proteins that are present in the muscle fiber in addition to actin and myosin. These are tropomyosin and troponin. When a signal from the motor neuron arrives at the synapse, tropomyosin is removed by Ca^{2+}-ions, which flood the muscle cell upon their release from the ER of the muscle cell (also called *s*arcoplasmatic *r*eticulum, *SR* for muscle-specific endoplasmic reticulum) through Ca^{2+}- and voltage-dependent Ca^{2+}-release channels (Ryanodine channels). The action potential from the motor neuron is transmitted at the neuromuscular junction by the neurotransmitter acetylcholine. Acetylcholine binds the nicotinic acetylcholine receptor, an ion channel gated by acetylcholine. By opening, it permits Na^+-ions to enter the muscle cell and increasing the membrane potential. This change in membrane potential activates voltage-dependent ion channels, further depolarizing the muscle cell membrane. When a threshold is reached, a Ca^{2+}-channel opens and provides the signal for Ryanodine channels of the sarcoplasmic reticulum to open and release Ca^{2+}-ions to induce muscle contractions (Fig. 4.9).

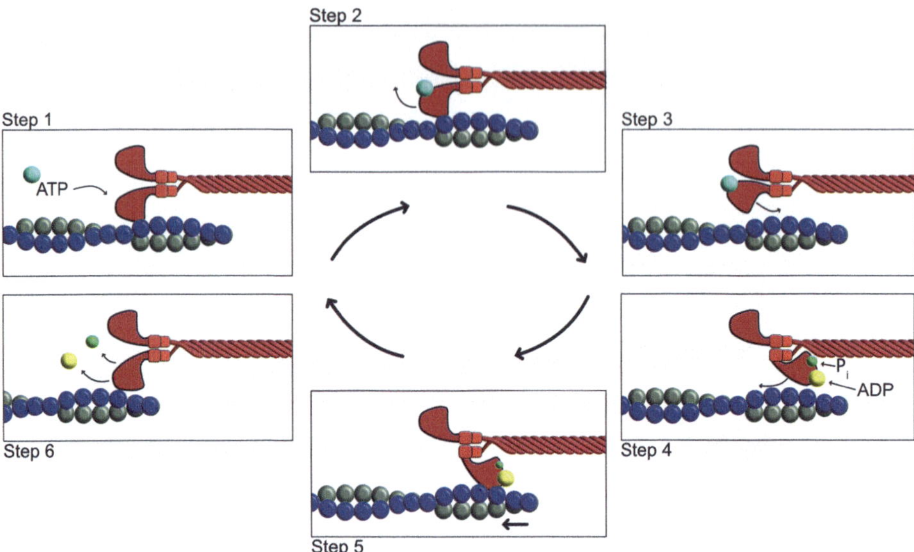

Fig. 4.8 Schematic representation of 6 steps of muscle contraction: ADP is removed from actin-bound myosin (Step 1); ATP (cyan sphere) binds to myosin ATP-binding site (Step 2), myosin is removed from actin, relaxation (Step 3), ATP is hydrolyzed to Pi (small green sphere) and ADP (yellow sphere), and Pi is not released from myosin. ADP-myosin experiences conformational change (Step 4), Pi gains free, myosin shifts actin filament, "power stroke," contraction (Step 5); ADP-bound myosin remains attached to actin (Step 6) until ADP is removed and exchanged for ATP (back to Step 1)

4.1.3 Macrofilaments (Microtubules)

Microtubules are built from heterodimers of two tubulin-molecules, α- and β-tubulin. Both subunits are able to bind one molecule of GTP each (Fig. 4.10). In the α-subunit, this GTP is "trapped," and it cannot be hydrolyzed. However, GTP of the β-subunit can be hydrolyzed to GDP and GDP can be exchanged for GTP. In this way, the β-subunit of the tubulin dimer changes between GDP- and GTP-bound states. This is important for the function of these dimers in the forming and dynamics of filaments. Simple human microtubules consist of 13 protofilaments, whereby protofilaments are strings of tubulin dimers. Thirteen of such strings form a tube (Fig. 4.10). Apart from simple tubes with 13 protofilaments, double tubes can also be formed, especially in cilia and flagella. The basal bodies of cilia and flagella as well as the centrioles of centrosomes are formed of triplet tubes (Fig. 4.10).

Microtubules have very specific functions in a cell. They are important for cell division by forming the spindle apparatus, which assists in separation of chromatids during mitosis (Fig. 4.11). Moreover, they stabilize axons and dendrites in nerve cells and provide transport cables for vesicle transport. These are especially important in nerve cells where vesicles often have to be transported from the core of the cell to synapses and membranes over long distances, e.g., in axons. In this case, motor proteins are associated with microtubules

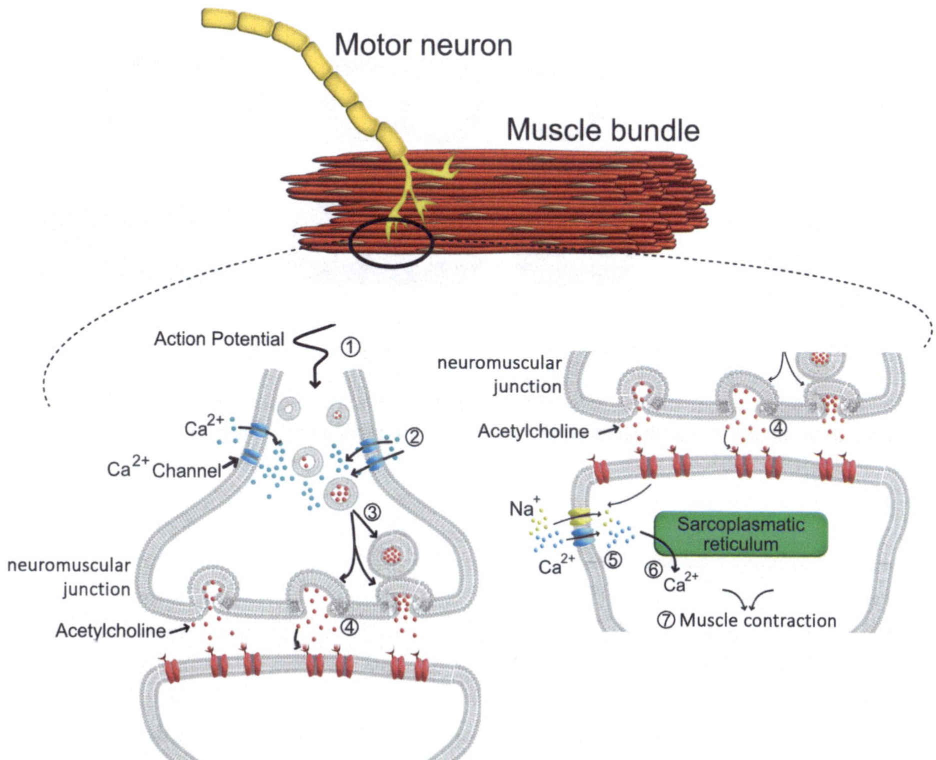

Fig. 4.9 Neuromuscular junction (oval black ring in upper panel): motor neurons (yellow) form synapse with muscle fiber cell (red with yellow nuclei); Lower panels: Enlargement of neuromuscular junction, and schematic representation of synapse on left-hand panel focuses on neuron: (1) action potential (black arrow) arrives, (2) voltage-gated Ca^{2+} channels open (cyan symbols in gray membrane), Ca^{2+} (cyan dots) influx, (3) acetylcholine-filled synaptic vesicles (red dots) fuse with synaptic membrane, (4) acetylcholine is released into synaptic cleft; Schematic representation of synapse on right-hand panel focuses on muscle cell: (5) acetylcholine binds nicotinic acetylcholine receptor ion channel (red symbols in membrane), channel opens and conducts Na^+ions (yellow dots and symbols in membrane), membrane depolarizes, voltage-gated Na^+-channels open, membrane depolarizes further until voltage-gated Ca^{2+}channel opens, and (6) Ca^{2+} influx triggers rhodamine receptor, a high conductance voltage and Ca^{2+}-gated Ca^{2+} channel in sarcoplasmic reticulum of muscle cell, Ca^{2+} is released (7) muscle fiber contracts

to carry the vesicles along microtubule arrays. Microtubules also form cilia and flagella (see later).

Each microtubule grows from the addition of more and more dimers of α- and β-tubulin. However, this growth has to be initiated from a stable site, in other words, microtubules have to be nucleated. The nucleation center is provided by a microtubule-organizing center (MTOC) involving yet another tubulin isoform, γ-tubulin. γ-tubulin is also a globular protein, which is associated into a spiral ring by binding to accessory proteins, resulting in a

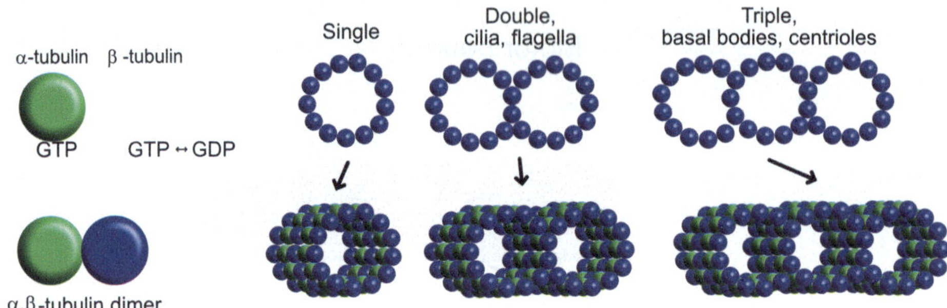

Fig. 4.10 Schematic representation of microtubules formed by protofilaments of α- and β-tubulin dimers (green and blue spheres). Protofilaments arrange into tubes as illustrated schematically in cross section for single microtubules consisting of 13 protofilaments, cilia with doublet of 2 incomplete microtubules made of 10 protofilamets each, and flagella with triplet of tubes made from 10 or 11 protofilaments each

so-called γ-tubulin ring complex (γTuRC). α- and β-tubulin dimers can be added onto this scaffold to build 13-protofilament microtubules (Fig. 4.12). The most prominent MTOC is the centrosome. It consists of two centrioles with a cylindrical shape formed by triplets of microtubules as shown in Fig. 4.10. The two centrioles are arranged perpendicular to each other. In cycling cells, they separate after mitosis and on each centriole, a new one is formed to achieve centrosome duplication before the cell divides again. The two centrosomes then move to opposite poles of the dividing cell and nucleate microtubules for the spindle apparatus. In resting cells, the centrosome can form the basal body of primary cilia. Cilia and flagella consist of double microtubules (Fig. 4.11). All cells are penetrated by a large number of microtubules, whereby many are nucleated at the centrosome, but others originate from cytoplasmic MTOCs.

Microtubules are dynamic. Not only do they grow from MTOCs, they also collapse at a certain length according to the processes they govern. The direction of growth is the (+) end, whereas the MTOC represents the (−) end. For growth, new tubulin dimers are added to the (+) end. When tubulin dimers are added, β-tubulin is always GTP-bound. The moment a new dimer with GTP-bound β-tubulin is added to the (+) end of the microtubule, the previous one starts GTP hydrolysis, but does not release P_i! It thus stores the energy of this hydrolysis step. The (+) end of microtubules is protected by capping proteins. When the cap is lost, the (+) end tubulin GTP is hydrolyzed, microtubules disassemble and the stored energy is released all of a sudden. In this way, microtubules can conduct "work" such as separating chromatids during mitosis. Such quick disassembly of microtubules is referred to as "catastrophy." The continuing process of growth and collapse of microtubules characterizes their "dynamic instability" (Mitchison & Kirschner, 1984) (Fig. 4.13).

Regulation of microtubule dynamics and remodeling (for instance, for spindle formation in mitosis) requires many additional factors, which are summarized as "Microtubule-associated proteins" (MAPs). This family, for instance, includes the Tau protein, which helps to form bundles of microtubules in axons. It is infamous for being present in

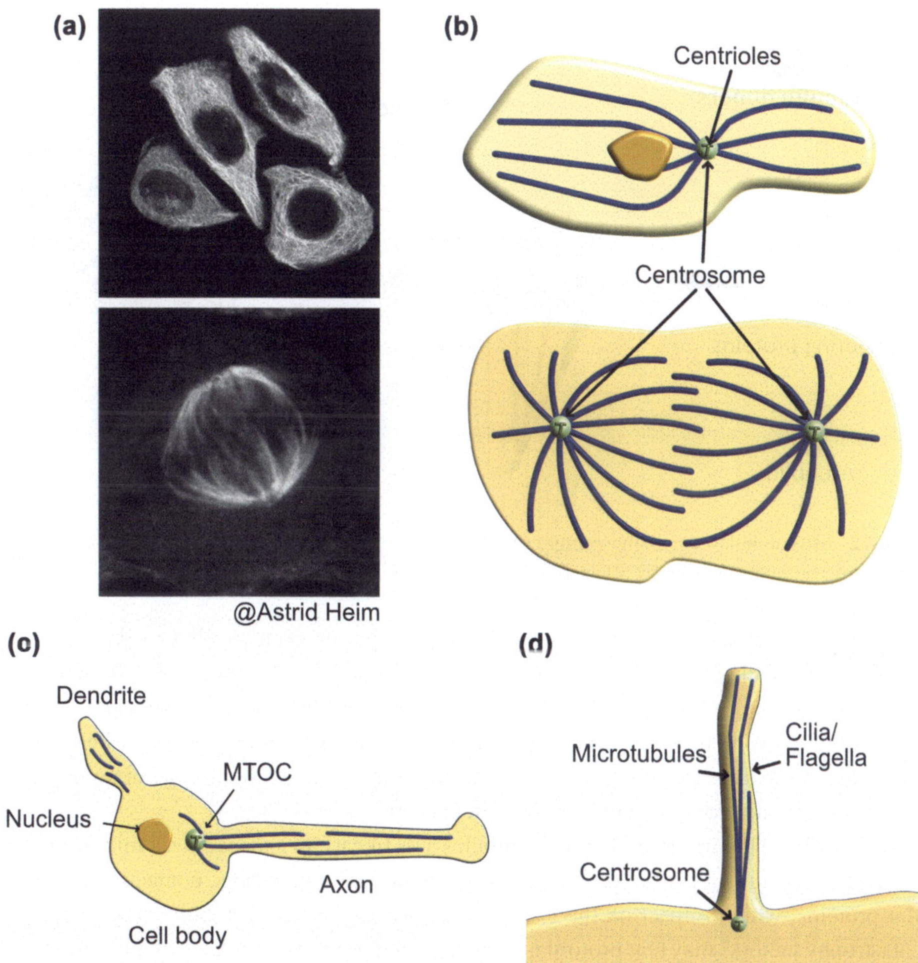

Fig. 4.11 Microtubules; (**a**) microtubules in HeLa cell stained with antitubulin antibody, upper cell in interphase, lower cell in late phase of mitosis when microtubules form mitotic spindle apparatus; (**b**) Schematic of microtubules in interphase with nucleus (orange) and centrosome (green, centrioles are indicated by T), and nucleus in anaphase with mitotic spindles of emerging daughter cells and centrosomes (green); (**c**) Nerve cell with axon, cell body and dendrites as indicated; mircotubules grow from MTOC; (**d**) Cilium with basal body and microtubule

so-called "neurofibrillary tangles" that are observed as plaques in the brains of patients with degenerative diseases including Alzheimer.

Like actin filaments, also microtubules can branch of existing microtubules. This involves a protein complex of six subunits, in *Drosophila* called Augmin, and of eight subunits called HAUS (homologous to augmin subunits) in humans. The name augmin was given because this complex augments the spindle apparatus by adding additional

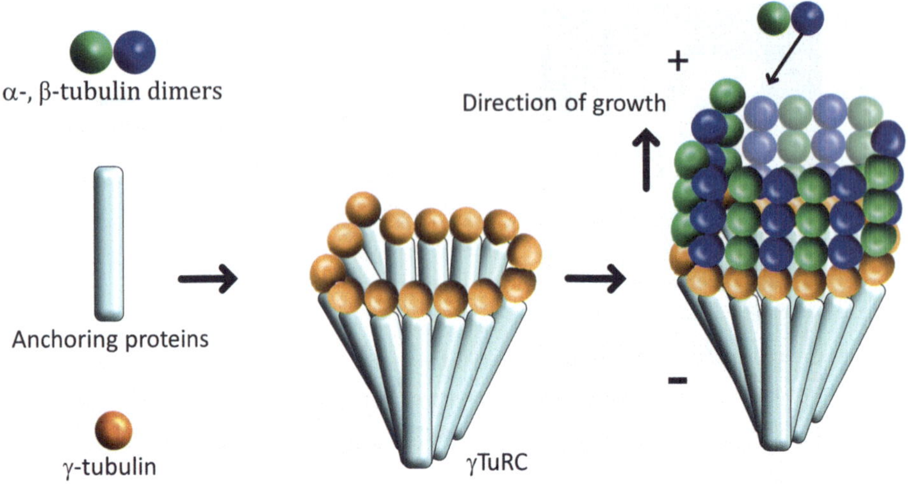

Fig. 4.12 Microtubule-organizing center consisting of anchoring proteins associated with γ–tubulin, and microtubules emerge by attachement of α, β-tubulin dimers

microtubules as branches, which run at shallow angles or even parallel to the existing spindle microtubules with the same polarity. Augmin attaches to tubulin at a branch site and recruits γ-tubulin for nucleation of new microtubules. Here, a γTuRC originates from pre-existing microtubules and functions as MTOC. In plants, mitosis and spindle formation occur in the absence of centrosomes and rely on conserved homologs of some subunits of the augmin complex (Prosser & Pelletier, 2017).

The γTuRC at the minus end of microtubules acts like a cap and stabilizes microtubules at the base. The (+) end, which carries the GTP-bound tubulin dimers, contacts many cellular proteins, which impact on the growth and shrinkage of microtubules. For instance, "catastrophy factors" may tear protofilaments apart and promote shrinkage of microtubules. Other factors can accelerate microtubule polymerization and yet another protein group binds to growing microtubules and can even attach them to certain organelles or to the chromosomes during mitosis. It is also possible that microtubules are broken. The proteins katanin and spastin, which are conserved throughout eukaryotic evolution, are ATP-dependent microtubule severing enzymes, which induce breakage of microtubules. They are essential for proper spindle dynamics. For instance, they can cut off microtubules from their γTuRC and thus induce depolarization of the spindle apparatus. They also cut off branched microtubules, which then have a "life" of their own. Moreover, katanin has important functions in the formation of cilia and flagella (Roll-Mecak & McNally, 2010).

An important factor for the polymerization of microtubules is the availability of free α- and β-tubulin dimers in the cytoplasm of the cell. Like with actin monomers, the cell has a pool of tubulin heterodimers. These can be brought to the (+) ends of microtubules by some proteins, which would intensify polymerization, or they can be kept away from microtubules by binding to certain proteins, e.g., stathmins. Stathmins thus reduce the

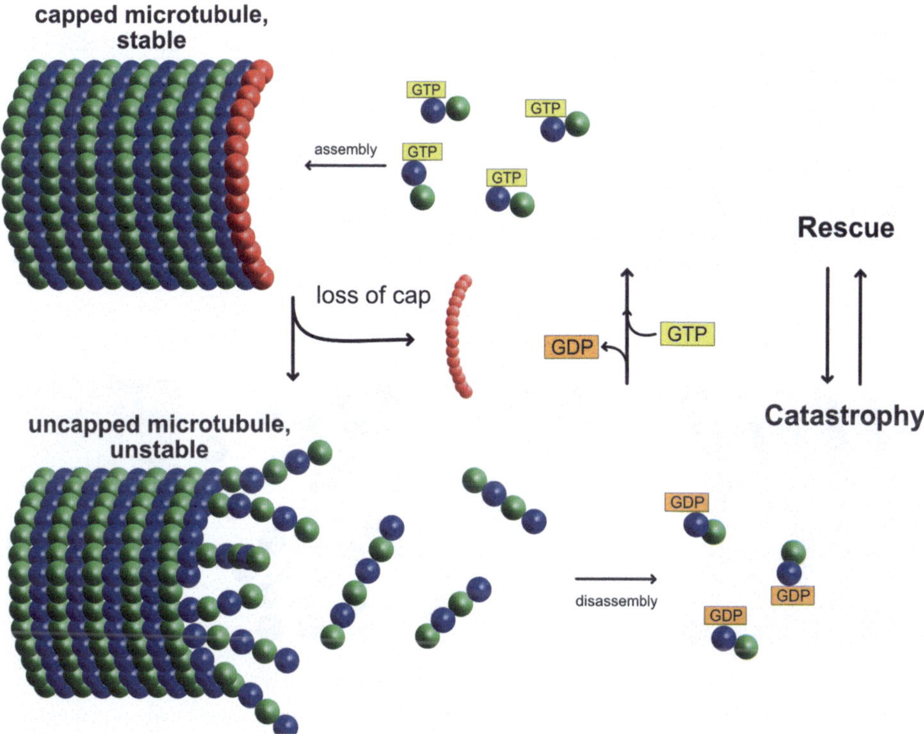

Fig. 4.13 Schematic illustration of dynamic instability of microtubules: Cap structure (red spheres) protects GTP bound to β-tubulin from hydrolysis, protofilaments form straight tube, and after loss of cap, GTP is hydrolyzed (catastrophy). Microtubules bend due to energy stored within protofilament, GDP- β-tubulin-α-tubulin dimers are released and exchange their GDP for GTP again; GTP-β-tubulin containing dimers are added to + end, microtubule growths (rescue)

availability of the building blocks for microtubules essentially leading to their destabilization, which is an important factor for regulating mitosis. Moreover, stathmins function in microtubules of motor neurons, where they are highly expressed. Tubulin heterodimer sequestration is inhibited by phosphorylation of stathmins, causing microtubule stabilization (Gagliardi et al., 2022).

Some drugs interfere with the "dynamic instability" of microtubules. One example is colchicine, the poison of an autumn flower, the meadow saffron *Colchicum automnale* (german *Herbstzeitlose*). It binds the tubulin heterodimer, preventing polymerization of new microtubules. Therefore, during cell division, a new spindle cannot be formed and this arrests the cell cycle in prometaphase. Another example is taxol, the poison of the yew tree *Taxus brevifolia* (german *Eibe*). Taxol stabilizes microtubules, they cannot depolymerize anymore and lose their "dynamic instability." This also makes it impossible to build a functional spindle apparatus in mitosis; therefore, cells treated with taxol, similarly to colchicine-treated cells, arrest in prometaphase, and the cell cycle is blocked (Fig. 4.14).

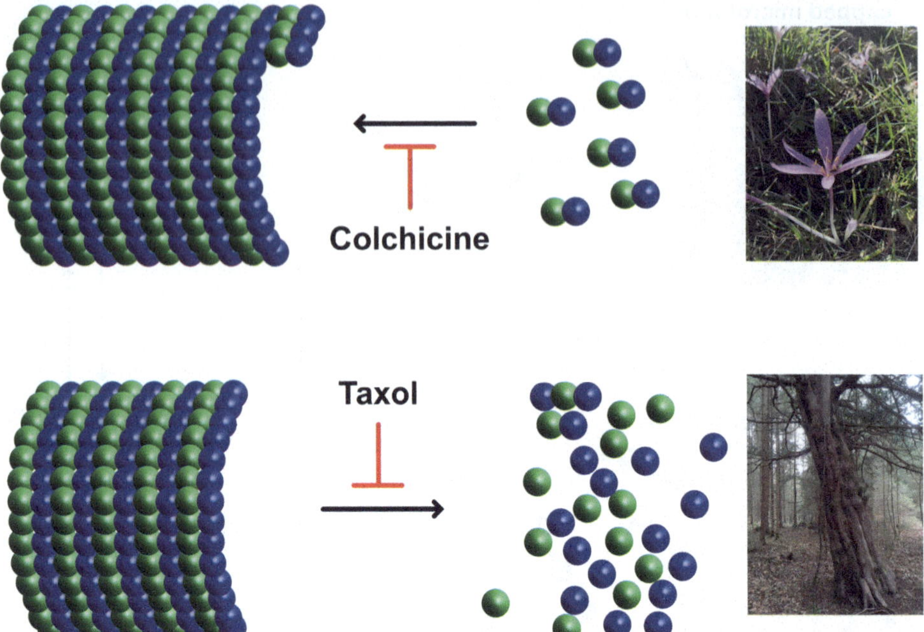

Fig. 4.14 Schematic illustration of colchicine and taxol activity on microtubules, colchicin from *Colchicum automnale* (eng: meadow saffron, german: *Herbstzeitlose*) as illustrated on photograph in the right-hand panel prevents microtubule assembly; taxol from *Taxus Baccata* (eng: Yew tree, german *Europäische Eibe*) as illustrated in photograph in the right-hand panel prevents disassembly of microtubules; Both therefore interfere with dynamic instability of microtubules

When vesicles or cell organelles are moved through the cells along microtubules, they use motor proteins, which attach to the vesicle and then "walk" along the microtubules with their cargo on the back (Fig. 4.15). For transport of vesicles toward the (+) end of microtubules, (e.g. along axons towards the synapse) in most cases, motor proteins of the kinesin family are employed, for transport towards the (−) end (e.g., late endosomes) motor proteins of the dynein family are used (Fig. 4.15). Both use ATP as an energy resource. The dynein family of motor proteins, however, has two branches. The first comprises cytoplasmic dyneins. Dynein 1, for instance is moving mRNA–protein complexes and cell organelles toward the (−) end of microtubules. The second branch comprises axonemal dyneins which are responsible for the beating of cilia and flagella that occurs by sliding of microtubules against each other.

4.1.4 Cilia and Flagella

Microtubules build a skeleton for cilia and flagella. These are structures on the cell surface in many protozoans and in most animal cells. They allow movement of fluids outside the cells, especially in epithelial cells. Flagella are also used for movement of whole organisms, e.g., protozoans or ciliates, or for movement of sperm cells. Cilia and flagella are

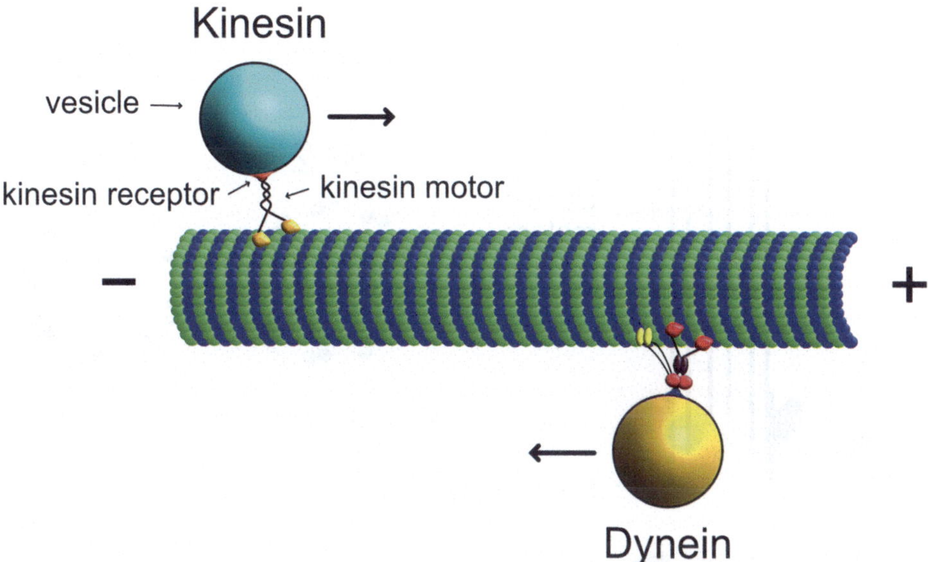

Fig. 4.15 Schematic illustration of kinesin motor attached to vesicle (blue sphere) and dynein attached to vesicle (yellow sphere), kinesin "walks" toward (+) end, dynein "walks" toward (−) end, for movement of both motors, head groups alternately attach to protofilament

formed inside the cell, they extrude underneath the plasma membrane, and they remain surrounded by it. The MTOC from which the cilial microtubules grow is represented by the basal body consisting of nine triplet centrioles from which the outer ring of microtubules grows. Two single microtubules can arise in the middle of the cilia forming the axoneme. In between the basal body and the axoneme is a transition zone. The individual microtubule doublets are connected by nexin proteins.

We distinguish between primary cilia and secondary cilia. Any cell that is not dividing produces a primary cilium, which plays a role for signal transduction (see later). Primary cilia are not actively moving. They are formed around a ring of nine doublet microtubules. Secondary cilia, in contrast, are actively moving. They also have a ring of 9 microtubule doublets; however, there are two additional microtubules in the middle. Between the inner and the outer microtubules, dynein motor proteins generate movements (Fig. 4.16), (Reiter & Leroux, 2017).

Motile and non-motile cilia are present in almost every organ in the human body. They include olfactory cilia in the nose, heart cilia, cilia of the oviduct, and respiratory cilia in the lung. The so-called outer segments of photoreceptor cells represent modified cilia (see Chap. 3).

Odontoblasts have prominent primary cilia. Their processes progress into the dentin matrix within dentinal tubules, which contain dentinal fluid. It is therefore assumed that odontoblast are involved in regulating the transfer of fluid, ions, and other molecules between dental tubules and dentin. Their primary cilia may have two different functions.

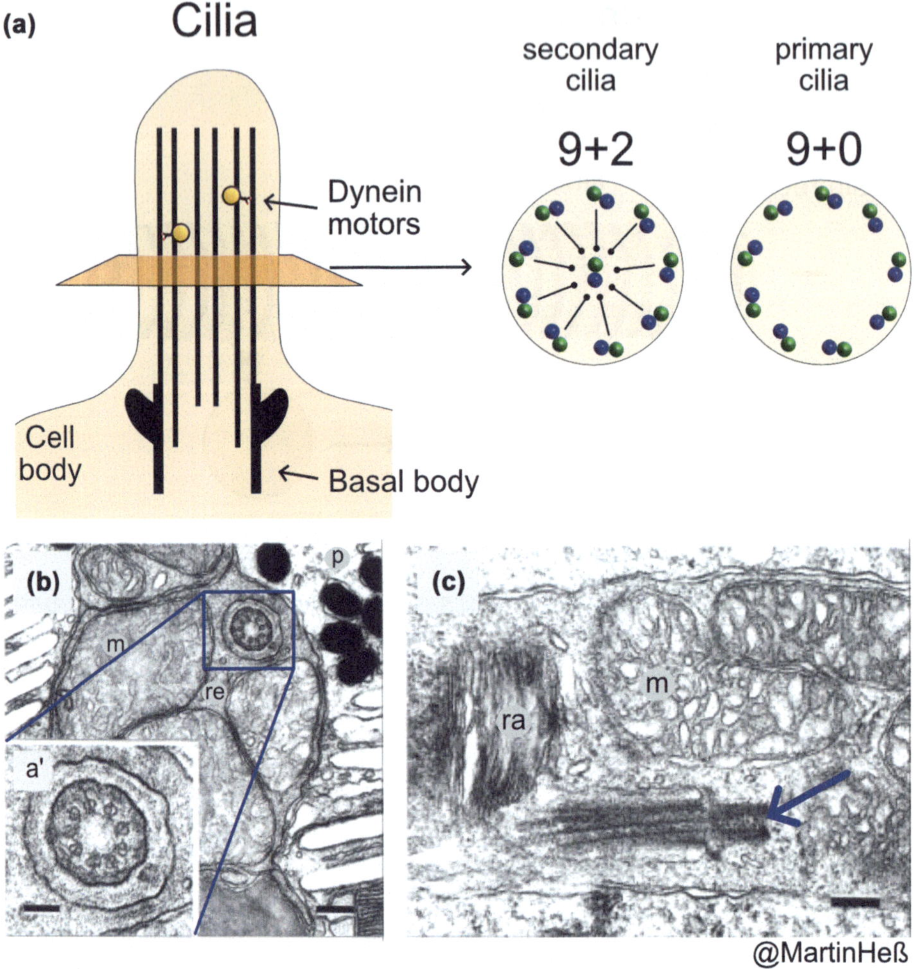

Fig. 4.16 (**a**) Schematic representations of cilia: Left-hand panel shows longitudinal section of motile cilium (secondary cilium) with microtubules (black lines) growing from basal body and dynein motor protein; Cross section of secondary cilium indicates 9 + 2 arrangement of doublet microtubules and position of dynein motor; Cross section of nonmotile primary cilium shows 9 + 0 arrangement without motor protein; (**b**) Electron micrograph of cross section of primary cilium in anchovy eye rod cell with typical 9 + 0 arrangement of microtubules, insert with enlargement of cilium, scale bar 100 nm; (**c**) radial section of cilium in anchovy eye rod cell (blue arrow), scale bar 250 nm; m mitochondrion, ra rod outer section @Martin Heß

During odontoblast differentiation and maturation, primary cilia probably receive developmental signals, which mediate direct cellular responses including dentin production, autophagy, and others. In addition, the odontoblast primary cilia could act as mechanosensory organelles. The dental flow of liquid in the dental tubules, while responding to environmental stimuli, such as cold, heat or acid, might cause deformation of primary cilia and this could activate mechanosensitive ion channels, which then transmit this information to nerve cells (Couve et al., 2013).

4.1.5 Intermediate Filaments

Intermediate filaments are constructed in a principally different way from macro- and microfilaments. Here, the filament-forming proteins display helical structures, which are intertwined to form stable coiled coils. A single intermediate filament protein consists of a globular head structure followed by a helical rod and a C-terminal globular tail. Two such molecules arrange in a parallel manner to form a coiled coil. These coiled coils then aggregate with another dimer in an antiparallel way to make a tetrameric protofibril. Four such protofibrils create a protofilament. The globular head structures are directed to the outside of these filaments where they regulate intermediate filament formation. The mature intermediate filament is eventually composed of protofilaments. Intermediate filaments are very tissue specific, intermediate filaments in skin, nails, and hair are keratins, desmins are the intermediate filaments of muscle cells, vimentin of fibroblasts and odontoblasts (box-fibroblast), and neurofilaments of neurons (Fig. 4.17).

Intermediate filaments are implicated in medical research. For instance, cytokeratins have been used as tumor markers due to their tissue specificity. Moreover, skin diseases are associated with keratins. Inheritable mutations in keratin genes have been shown to be responsible for *Epidermolysis bullosa*. In this condition, the contacts of epidermal cells with the basal lamina are disturbed, which leads to a detachment of the skin from the basal lamina and the formation of skin blisters.

Additionally, the nuclear lamina, a fibrous protein net covering the inner surface of the nucleus, consists of special intermediate filaments, assembled by lamin proteins. The nuclear lamina falls apart during mitosis of a parent cell. This is caused by phosphorylation of the individual lamin proteins at the first stage of mitosis (see Chap. 5). At the end of mitosis, lamins are dephosphorylated and reassemble into two nuclear laminas surrounding the new daughter nuclei.

Mutations in lamins have been found responsible for an extremely rare disease called Hutchison–Gilbert Progeria, which has been brought to public attention by a Hollywood movie featuring Robin Williams. Patients carrying this mutation age very quickly, and the precise mechanisms causing this defect are not known. However, this is by far not the only heritable disease caused by mutations in lamin proteins. The nuclear lamina is involved in many regulatory processes in the nucleus. These include chromatin organization, transcriptional and splice regulation and others, all of which impact cell division and differentiation processes. Compromising these functions can thus cause developmental defects.

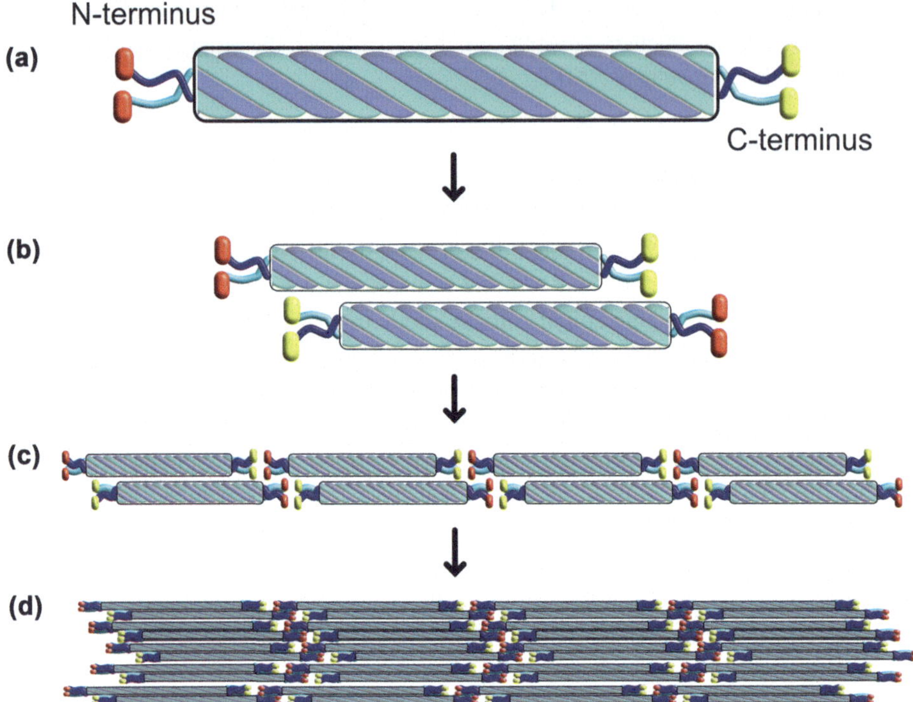

Fig. 4.17 Schematic representation of intermediate filaments: (**a**) 2 molecules (purple and cyan) of intermediate filament proteins with long helical middle structure and globular parts at the N- and C-termini are arranged in an antiparallel manner forming a coiled-coil structure in the helical parts; (**b**) Coiled-coil dimers are arranged in a parallel manner to form tetramers (protofilament); (**c**) Tetramers are arranged to form protofibrils; (**d**) Protofibrils are packed into filaments

4.2 Cellular Contacts, Cell Adhesion, and Extracellular Matrix

4.2.1 Cellular Contacts and Cell Adhesion

Cells in our body are organized in tissues and organs. Epithelia are lining vessels and cavities such as the arteria and veins, the stomach, intestine, and lung. Our skin is composed of epithelia, which form several layers. Epithelia are constituted from single epithelial cells. They are usually polar cells, meaning that the apical side that points to the inside of the cavity they line, or to the outside of the body, differs in structure and function from the basal side that points toward the part of the body or organ, which the epithelium is covering (Fig. 4.18).

The cells forming an epithelium are connected to each other. There are four main types of connections, which have several important functions (Fig. 4.18). They are visible in electron micrographs and early images were presented in a paper by Farquhar and Palade in 1963 (Farquhar & Palade, 1963). This work shows the junction between two epithelial cells of the intestinal mucosa of a rat with three types of junctional complexes. First, a tight connection between the epithelial cells sealing the epithelium from the environment

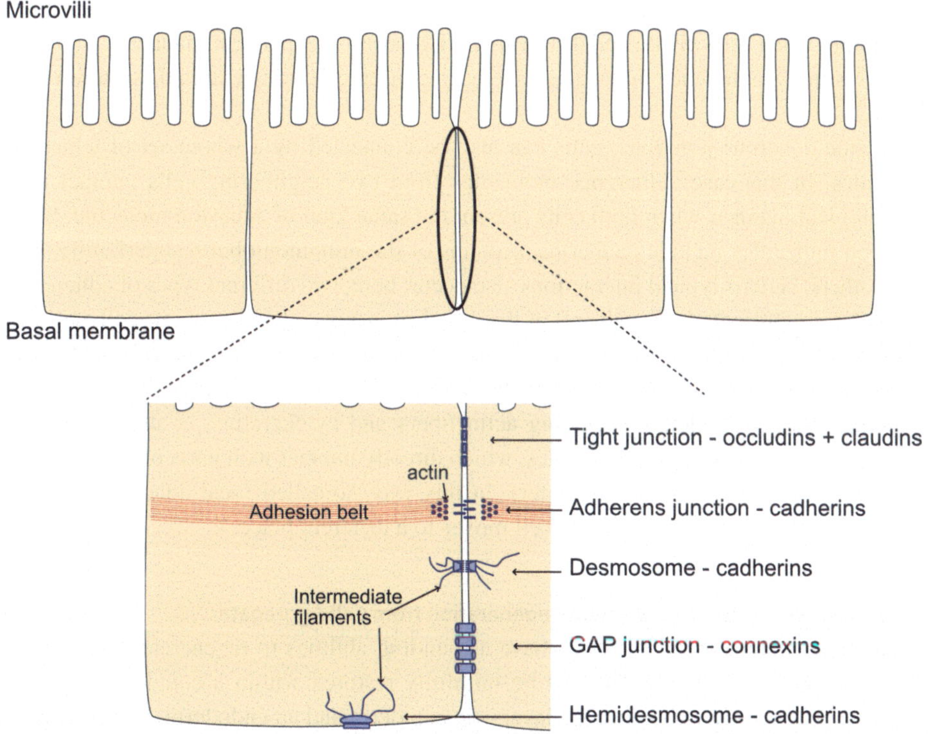

Fig. 4.18 Schematic representation of epithelial cell layer with microvilli at apical side and basal membrane as indicated at basal side. Lateral cell connection between two epithelial cells is magnified and cellular junctions are indicated by blue symbols including tight junctions, adherens junctions, desmosomes, GAP junctions, and hemisdesmosomes connecting the cell with the extracellular matrix at the basal membrane

is provided by tight junctions. This allows control of ion and water flow through the epithelia and provides mechanical stability. Tight junctions are composed of occludins and claudins, both four-times-membrane-spanning proteins. Two claudins (claudins 3 and 4), preferentially found in the intestine, are receptors for bacterial enterotoxins. When during a *Clostridium* infection such enterotoxins bind to claudins, the intestinal epithelium is severely harmed, leading to diarrhea.

A second type of junction are adherens junctions. Adherens junctions are located directly underneath tight junctions (Fig. 4.18). They are cell type-specific and their major components are cadherins. Cadherins represent a large family of single-pass transmembrane proteins. They are glycosylated at their extracellular domains and cadherins of a specific cell type usually only make contacts with the same cell type, strictly in a Ca^{2+}-dependent manner. For instance, N-cadherins occur between cells in nervous tissue, whereas E-cadherins occur in epithelial tissue. The intracellular parts of cadherins are connected with actin fibers via anchoring proteins. In this way, a so-called adhesion belt is formed (see Fig. 4.18). Cellular signaling pathways regulate the shape and stability of the epithelium by interacting with adherence molecules. This is especially important for embryonal development or developmental processes in adults, for instance, for the constant renewal of the intestinal epithelium, when cells divide and differentiate in a position-dependent manner.

Interestingly, adherence molecules have been conserved through animal evolution (see Box 4.2).

Outside adherens junctions, cells can also be connected by a whole set of adhesion molecules. In this case, adherence molecules from two neighboring cells interact in a homo-typical manner, when both cells present the same kind of adhesion molecule. Such adhesion molecules include cadherins, proteins of the immunoglobulin superfamily, and many others. Hetero-typical interactions also occur; here, two different types of adherence molecules are presented on each cell of the pair. For instance, lectins on one cell can specifically interact with glycosylated parts of certain glycoproteins on the cell next door. Moreover, adherence molecules can be connected with the extracellular matrix. Therefore, the cytoskeleton, including actin fibers and cytokeratins, is attached to the intracellular part of adhesion molecules, which directly interact with parts of the extracellular matrix, e.g., fibronectin (see later). In this way, so-called focal adhesions can be created, but also destroyed when the cell moves to a different place.

Box 4.2 Cell Sorting During Hydra Regeneration from Cell Aggregates

The fresh water polyps of *Hydra* have astounding abilities to regenerate. One can remove any body part and it will be faithfully restored within a couple of days. *Hydra* polyps have two epithelial layers, an ectoderm and an endoderm. Polyps can be dissociated into single cells, whereby ectodermal and endodermal cells are mixed. When all cells are pipetted together, they will quickly sort themselves out and form a ball with the former ectodermal cells connecting to each other and forming an outside layer and endodermal cells doing the same and forming the inner layer. From this ball, new polyp heads and feet are formed and eventually new hole polyps arise. Sorting of the two epithelial cell layers is possible because specific cadherin molecules make new cell type-specific connections between the cells of each type. This requires Ca^{2+}-ions. Therefore, if you try to do such an aggregation experiments in the absence of Ca^{2+}-ions, it inevitably fails.

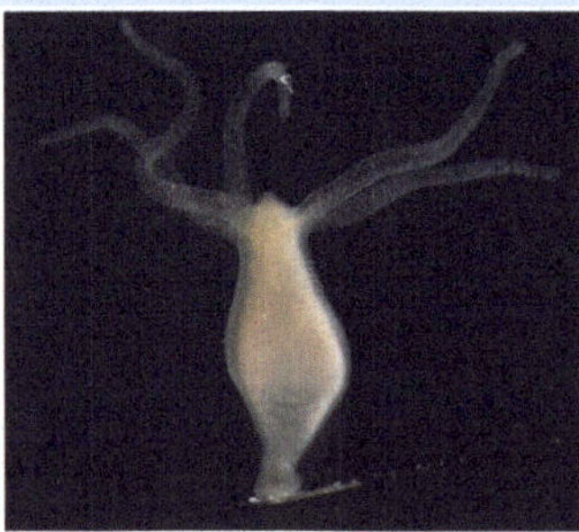

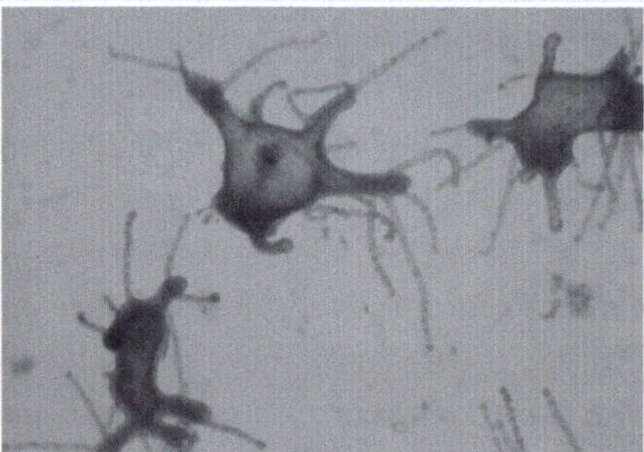

This picture illustrates further developed aggregates of dissociated cells of hydra polyps (left-hand side).

The third type of cellular epithelial junctions is provided by desmosomes (Fig. 4.18). Desmosomes are also formed by conglomerations of cadherins, which are anchored to intermediate filaments at their intracellular side. This is important for the stability of the epithelium. A special case are hemidesmosomes. These structures connect the intermediate filaments inside the cell with the extracellular matrix. Mutations in cytokeratins can interfere with desmosome and hemidesmosome formation and cause severe hereditary skin diseases, such as epidermolysis bullosa, the name indicating that the epidermis is "dissolved," visible as blisters on the skin. Hemisdesmosomes are similarly constituted by cadherins that are connected to intermediate filaments, and the extracellular part of the cadherin is then attached to the extracellular matrix. Hemisdesmosomes are found at the basal side of epithelial cells where the extracellular matrix forms a basal lamina.

Finally, there are also cell junctions, which make real open pores and thus allow direct flow of small molecules or ions in lateral direction between epithelial cells. These are GAP junctions (Fig. 4.18). Gap junctions are therefore communication contacts. They are constructed by connexins, another family of four-times-membrane-spanning proteins. Four connexins molecules assemble a connexon, a channel in the epithelial cell membrane (Fig. 4.19). When two such channels from neighboring cells join, a Gap junction is built. The intracellular parts of the connexin molecules are regulated by adaptor proteins, phosphorylation, and other mechanisms, which allows regulation of the pore size. In general, molecules up to a molecular mass of 1000 Da can pass a Gap junction. Moreover, Gap junctions provide electrical synapses, e.g., in the heart, the retina, and the brain because the open channels allow membrane potential changes to be passed between cells without a chemical synapse.

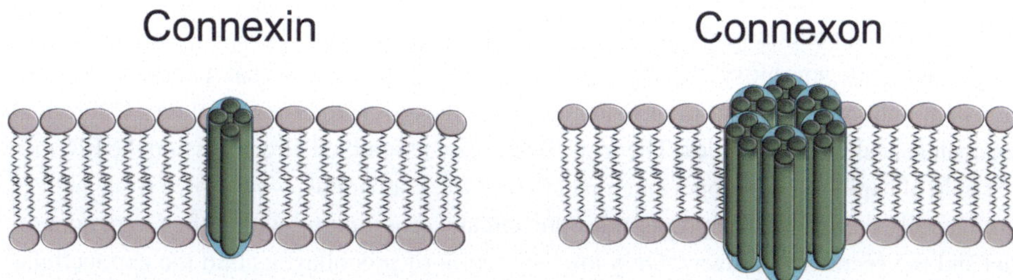

Fig. 4.19 Schematic representation of connexin (4 times membrane spanning protein) and connexon composed of 6 molecules of connexin as indicated, connexons from 2 adjacent cells can be joined and form a GAP junction

4.2.2 Extracellular Matrix

The human body as any other organism is made from cells. However, not every molecule in our body, not even every macromolecule is present inside these cells. We have teeth, joints, and bones, for instance. These are part of the extracellular matrix, which is synthesized by cells, but then secreted into spaces outside. This extracellular matrix has many functions. It is important for communication between the cells, it supports mechanical properties of the body, and it protects the surface tissue of the organism from the environment. All extracellular matrix in animals has four major components including collagen, hyaluronic acid, proteoglycans, and anchoring proteins. These components are produced by the cells and each cell type produces its own special extracellular matrix. In epithelial cells, the extracellular matrix forms a thin layer called basal lamina at the basal side of the epithelium. The side oriented toward the outside usually has a very pronounced glycocalyx including quite diverse and cell type-specific carbohydrate chains, which have also been named “sugar coats” (Gabius et al., 2022). In connective tissue, such as bones, teeth, or joints, the matrix can become very large, which requires a high production rate of the cells within those tissues. In the following, the basic components of extracellular matrix are described.

Collagen is the most abundant protein in our bodies and makes about one quarter of our protein mass. It has a special structure. Its amino acid chain is very regular with a glycine residue at every third position, thus displaying a primary sequence of (X-Y-Gly)n. X and Y are often represented by lysine and proline residues. This sequence forms a helix. In contrast to the usual α-helix, due to the small amino acid glycine in every third position, this helix makes a turn every 3 amino acids and this allows three helices (or three collagen monomers) to intertwine, ending in a triple helix with an extraordinary stability. The glycine residues are arranged in the middle of the triple helix. These triple helices form fibrils and are the main components of connective tissue (Fig. 4.20).

During collagen biosynthesis, monomers, which possess signal peptides, are co-translationally inserted into the ER as pre-pro-collagens with N-terminal and C-terminal extensions, called propeptides. The triple helices are assembled already inside the ER with the help of hsp47, a chaperone, and travel to the Golgi apparatus, where three triple helices are laterally compiled. The propeptides prevent any crosslinking within the Golgi apparatus and the secretory pathway. Only after secretion of procollagens into the extracellular space, propeptides are removed and the collagens assemble laterally and longitudinally into multimers, which are chemically crosslinked and eventually form large and stable fibrils (Fig. 4.21). These fibrils show cross-striations in EM images, which indicate the length of the collagen triple helices of 67 nm. Collagens are also present in biomineralized extracellular structures such as bones and teeth. For crosslinking, oxygen plays an important role. Lysine residues at the ends of the helices are hydroxylated within the secretory compartments by prolyl- and lysyl hydrogenases, both Fe(II) and 2-oxoglutarate-dependent

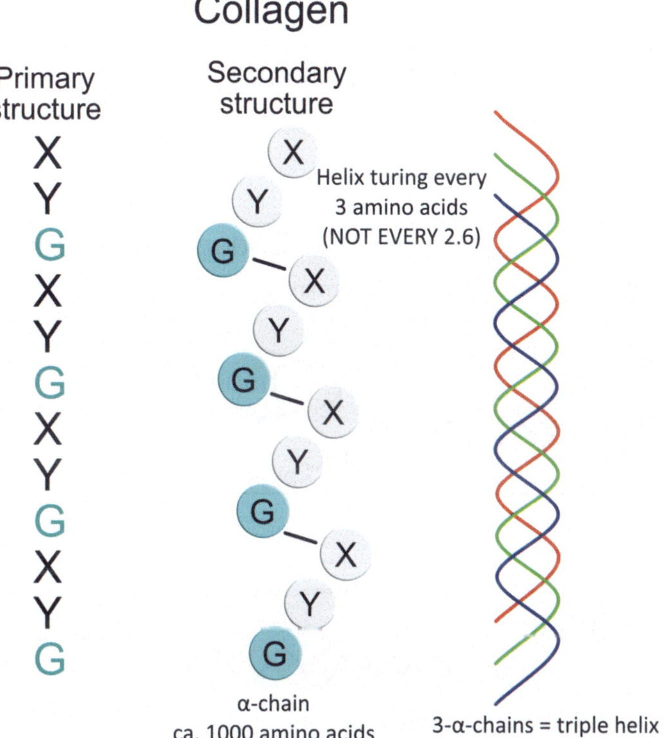

Fig. 4.20 Schematic representation of collagen primary structure with amino acid sequence X (any of the 20 amino acids); Y (single letter code for tyrosine); G (glycine), secondary structure forming α-helix with 3 amino acids per turn and tertiary structure forming triple helix by intertwining of 3 helical collagen molecules

enzymes. This reaction also requires ascorbic acid (Vitamin C), for efficient activity. Moreover, lysyl residues are oxidized by lysyloxidase to form aldehydes. Aldehydes and hydroxyl groups of different α-helices form a covalent bond between individual helices of the triple helix, or between different triple helices, to stabilize the collagen fibrils. In Vitamin C-deficient situations, these crosslinks cannot form, which delays the replacement of collagen fibrils and leads to loss of teeth, problems with wound healing, and destabilization of blood vessels. These symptoms are known from sailors that did not receive any fresh fruit or vegetables while roaming the oceans in times previous to the nineteenth century, when this connection between Vitamin C and scurvy was discovered and action was taken to improve the diet of sailors. Aging also has an effect on the formation of collagen, especially on the crosslinking. The activity of the enzymes involved changes enormously with age. This leads to progressive wrinkling of the skin (Birch, 2018).

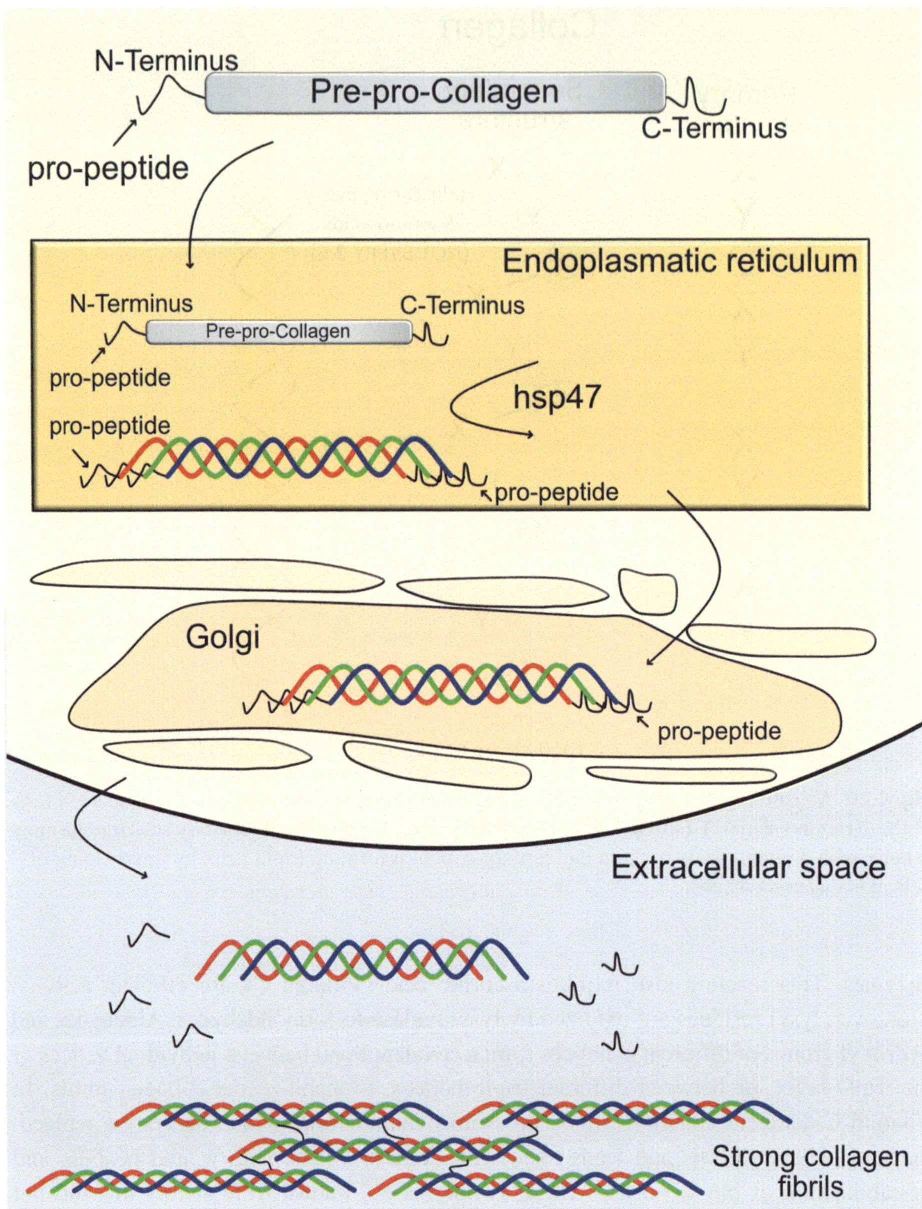

Fig. 4.21 Schematic representation of collagen biosynthesis: collagen is co-translationally transported into ER, helices are folded into triple helix with help of chaperone hsp47, triple helix (pre-pro-collagen) is formed inside the ER where lysine and prolin residues are hydroxylated, pre-pro-collagen is transported to the Golgi and glycosylated, secreted into the extracellular space and N- and C-terminal amino acids are cleaved of, and strong collagen fibers are formed in extracellular space by crosslinking of pre-pro-collagen fibrils

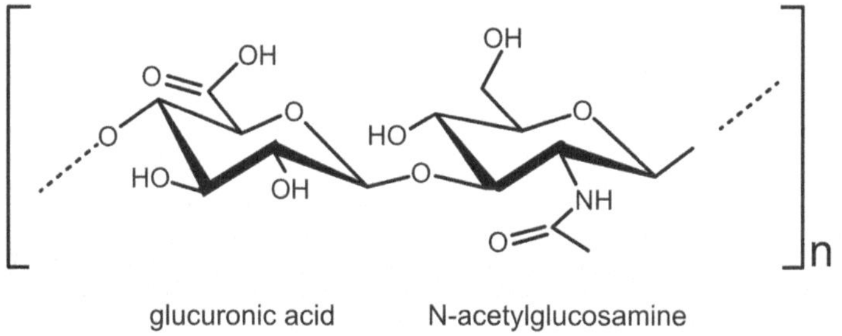

Fig. 4.22 Molecular structure of the polysaccharide hyaluronic acid consisting of repetitions of the disaccharide glucuronic acid - N-acetylglucosamine

Hyaluronic acid is a polysaccharide consisting of repeats of the disaccharide glucuronic acid and N-acetylglucosamine (Fig. 4.22). It reaches a molecular mass of 8 million Da (compare globular proteins with 50,000 Da). Due to its hydroxyl groups, it binds enormous amounts of water and reaches a gel-like consistency as we find it in joints and cartilage.

Proteoglycans are proteins which covalently bind chains of glycosaminoglycans. A linker tetrasaccharide consisting of xylose, two molecules of galactose and glucuronic acid- is attached to serine residues of the protein part of the structure. The glucuronic acid then links up with larger polysaccharide chains (Fig. 4.23). For example, the protein decorin with a molecular mass of 40,000 is the core protein for the glucosaminoglycan aggrecan, which reaches a molecular mass of three million. Anchoring proteins lock the extracellular matrix to the cell membrane. An example are integrins, transmembrane proteins, which play an important role in transmitting signals between the cell interior and the extracellular matrix. (Fig. 4.24). A very specific collagen, collagen XVII, can also function as anchor protein because it is directly inserted into the plasma membrane. The ECM protein laminin bridges these transmembrane protein anchors with collagens. In this way, a direct communication between extracellular matrix and cell interior becomes possible.

4.2.3 Enamel and Dentin

For tooth formation, the extracellular matrix plays a major role. The major compound of mineralized dental tissue is dentin. Dentin is covered by enamel at the crown of the tooth and by cementum at its base.

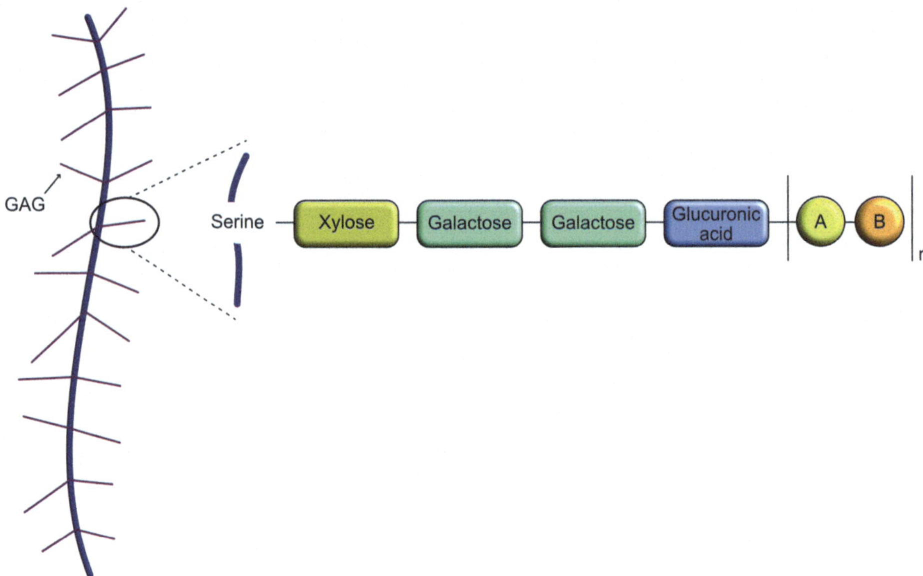

Fig. 4.23 Schematic representation of proteoglycans, composed of a long core protein (blue line) covalently bound to glycosaminoglycans, and at serine residues, bonds are formed with linker tetrasaccharide (Xylose–galactose–galactose–glucuronic acid) followed by repetitions of a specific disaccharide (here shown as yellow and orange spheres A and B)

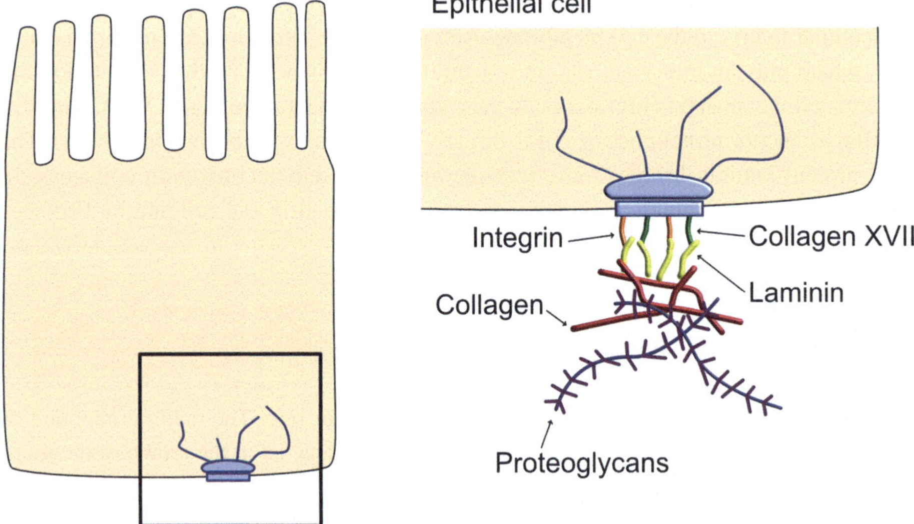

Fig. 4.24 Schematic illustration of hemidesmosome anchoring the basal membrane to the extracellular matrix; Left-hand panel shows schematic of epithelial cell with basal hemisdesmosome (blue); Right-hand panel shows magnified hemidesmosome with integrins (transmembrane proteins, green) and the transmembrane collagen XVII (BP180, orange) attached to laminin anchor protein (yellow), which connects it to collagen fibers (red) and to proteoglycans (purple) of extracellular space

Enamel is the hardest material in the human body. Its synthesis is initiated during embryonic tooth development by ameloblasts. These cells are lost when the tooth breaks through into the mouth. The mineral content of enamel amounts to 96%. On the organic side, it contains the proteins enamelin and amylogenin, providing the scaffold for mineralization. The second dental cell type are odontoblasts. These cells synthesize extracellular matrix proteins and are responsible for their secretion, transport, and mineralization (Fig. 1.7). In their mature form, they are polarized cells, which are aligned at the surface of the dental pulp in a palisade-like manner rectangular to the basal membrane. They have a cell body with the ER, Golgi apparatus, lysosomes, secretory vesicles, and the cytoskeleton, involving microtubules, actin filaments, and vimentin and nestin containing intermediate filaments. The cell bodies form junctions with their neighbors including desmosomes, Gap junctions and sometimes tight junctions, which provide a permeability barrier between dental pulp and dentin (Goldberg et al., 2011). From the odontoblast cell body long processes reach into the dentin tubules. These support secretion and mineralization of ECM proteins. Dentin occurs in three forms in the tooth. These are the mantle dentin underneath the enamel crown, the intertubular dentin between the tubes where the odontoblast processes lie, and the peritubular dentin close to the odontoblast. The outer layer mantle dentin in a mature tooth is supported by matrix vesicles which bud from odontoblast processes. The intertubular dentin constitutes the major dentin part of the tooth. It is formed on ECM proteins secreted by the odontoblast. These include collagen, fibronectin, glucosaminoglycans, and non-collagenous phosphorylated proteins. The diameter of collagen fibrils increases from ca. 20 nm near the odontoblast cell body to 40 nm in the intertubular dentin and 55 to 75 nm in the mantle dentin. By lateral aggregation of collagen fibrils, a network is built. Mineralization occurs on this collagen scaffold. The first step of dentin mineralization is the deposition of ECM components fibronectin and then collagen into the ECM. Then non-collagenous ECM components are phosphorylated by a kinase (casein kinase) and sulfated by sulfatases. Eventually, Ca^{2+}-ions interact with acidic residues of ECM proteins and combine with phosphate. This leads to the formation of the hydroxyapatite crystals of dentin and their deposition within the collagen template (for review, see Goldberg et al. (2011)). The third layer of dentin is the peritubular dentin. It forms a gel-like network from phospho-, glyco-, and lipoproteins and does not contain collagen. It is assumed that it is supported by passive deposition of blood serum-derived components.

Test Yourself

1. What are microfilaments made of?
2. Which nucleotide delivers energy for actin fibers?
3. Are actin filaments polarized?
4. Which proteins play an important role for actin dynamics?

5. How does muscle contraction in skeletal muscles work?
6. Name two cellular structures that are formed by microtubules.
7. Do microtubules display polarity?
8. The tubulin molecule is of about the same size as the actin molecule (ca. 50 kDa). Why do microtubules have a much higher diameter?
9. Which motor proteins are acting on microtubules?
10. How big are intermediate filaments in comparison to other components of the cytoskeleton?
11. What is the nuclear lamina?
12. Name four components of the extracellular matrix.
13. Name three body structures made from extracellular matrix.
14. How is the ECM connected with the cell membrane?
15. What is the function of tight junctions between lateral membranes of epithelial cells?
16. What cell–cell connections allow passage of small molecules between epithelial cells?
17. What cell–cell connections are linked to actin?
18. Which cell–cell connections are linked to intermediate filaments?
19. Name one important adhesion molecule.
20. What are the molecular components of enamel and dentin?

References

Birch, H. L. (2018). Extracellular matrix and ageing. *Sub-Cellular Biochemistry, 90*, 169–190.

Bugyi, B., & Kellermayer, M. (2020). The discovery of actin: "To see what everyone else has seen, and to think what nobody has thought". *Journal of Muscle Research and Cell Motility, 41*(1), 3–9.

Cooke, R. (2004). The sliding filament model: 1972–2004. *The Journal of General Physiology, 123*(6), 643–656.

Couve, E., Osorio, R., & Schmachtenberg, O. (2013). The amazing odontoblast: Activity, autophagy, and aging. *Journal of Dental Research, 92*(9), 765–772.

Farquhar, M. G., & Palade, G. E. (1963). Junctional complexes in various epithelia. *The Journal of Cell Biology, 17*(2), 375–412.

Gabius, H. J., Cudic, M., Diercks, T., Kaltner, H., Kopitz, J., Mayo, K. H., Murphy, P. V., Oscarson, S., Roy, R., Schedlbauer, A., Toegel, S., & Romero, A. (2022). What is the sugar code? *Chembiochem, 23*(13), e202100327.

Gagliardi, D., Pagliari, E., Meneri, M., Melzi, V., Rizzo, F., Comi, G. P., Corti, S., Taiana, M., & Nizzardo, M. (2022). Stathmins and motor neuron diseases: Pathophysiology and therapeutic targets. *Biomedicine, 10*(3), 711.

Goldberg, M., Kulkarni, A. B., Young, M., & Boskey, A. (2011). Dentin: Structure, composition and mineralization. *Frontiers in Bioscience (Elite Edition), 3*(2), 711–735.

Kühne. (1864). *Untersuchungen über das Protoplasma und die Contractilität*. Wilhelm Engelmann.

Mitchison, T., & Kirschner, M. (1984). Dynamic instability of microtubule growth. *Nature, 312*(5991), 237–242.

Pollard, T. D., & Goldman, R. D. (2018). Overview of the cytoskeleton from an evolutionary perspective. *Cold Spring Harbor Perspectives in Biology, 10*(7), a030288.

Prosser, S. L., & Pelletier, L. (2017). Mitotic spindle assembly in animal cells: A fine balancing act. *Nature Reviews. Molecular Cell Biology, 18*(3), 187–201.

Reiter, J. F., & Leroux, M. R. (2017). Genes and molecular pathways underpinning ciliopathies. *Nature Reviews. Molecular Cell Biology, 18*(9), 533–547.

Roll-Mecak, A., & McNally, F. J. (2010). Microtubule-severing enzymes. *Current Opinion in Cell Biology, 22*(1), 96–103.

Szent-Györgyi, A. (1963). Lost in the twentieth century. *Annual Review of Biochemistry, 32*, 1–14.

5 Cellular Homeostasis: Cell Division, Cell Death, and Virus Infection

What You Will Learn in This Chapter

Cells arise from cell divisions of parent cells. During sexual reproduction, a new organism develops from one single cell, the fertilized egg. Cell division occurs in cycles where the DNA is doubled in S phase by DNA replication and then partitioned onto the two daughter cells during mitosis. In this chapter, we discuss the mechanisms of mitosis and cell cycle control in detail. Moreover, we look at the signal transduction pathway by which growth factors regulate cell proliferation and differentiation.

In order to maintain homeostasis of all cell types in a multicellular organism, also cell death (apoptosis) is regulated. A cellular suicide program is initiated by cells when they either are damaged or when they have fulfilled their functions and are not needed anymore. This program is initiated by extracellular signals or by intracellular mechanisms, which often involve the release of mitochondrial proapoptotic factors. These signals activate caspases, which then orchestrate the demise of the cell.

Finally, we discuss the structure and function of viruses. Viruses carry genetic information as RNA or DNA within an envelope or a capsule. They have to enter a host cell in order to replicate because they do not have ribosomes and thus cannot manufacture the proteins which are encoded in their nucleic acids.

5.1 Cell Division and Cell Cycle

Cells divide and as far as we know, this is the only way how new cells can arise. Cells cannot develop de novo from organic material, as it had originally been suggested in the nineteenth century (see Chap. 1). We know now that cell divisions are responsible for the reproduction of single-celled organisms, such as bacteria, protozoans, or budding yeast.

A. Böttger, J. Moneer, *Basic Cell Biology for Dentists and Medical Students*, https://doi.org/10.1007/978-3-032-03636-0_5

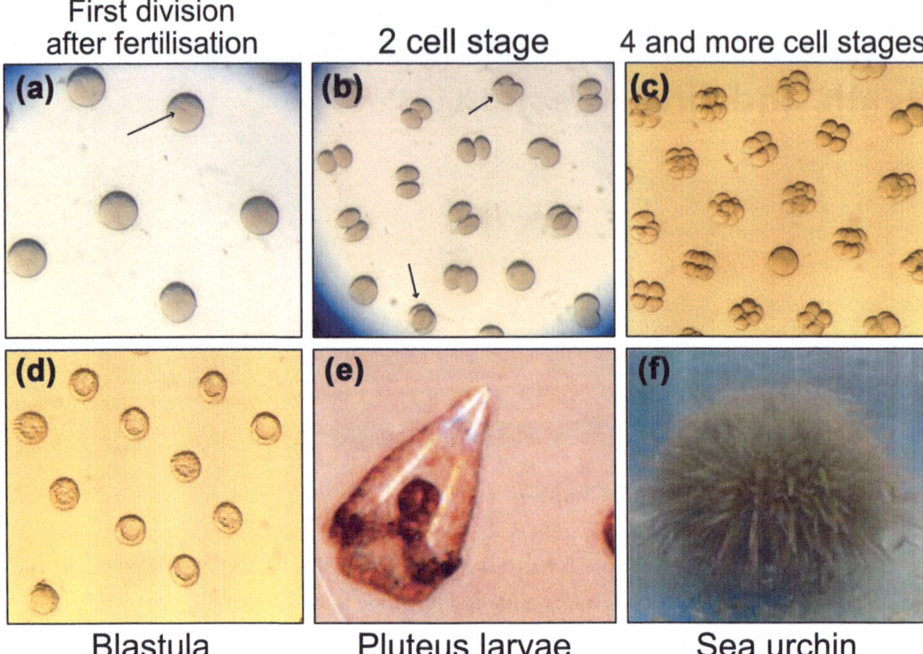

Fig. 5.1 Cell divisions during sea urchin development: (**a**) Fertilized eggs (zygotes) show small "cleavage furrow" (black arrow), indicating the beginning of cytokinesis at the end of first mitosis; (**b**) Two cell stages and some four cell stages (black arrow) after second cell division of zygote; (**c**) Second and following cell divisions have occurred; (**d**) Blastula stages; (**e**) Pluteus larvae developing after 3 days; (**f**) Mature sea urchin taken from North sea dredged material for comparison. (Photographs taken by sea urchin lab course 2014, Ludwig-Maximilians-University Munich)

Cell divisions are necessary for asexual reproduction of plants or budding of *Hydra* polyps. And most importantly, every multicellular organism capable of sexual reproduction, be it fungus, plant, or animal, arises from a single cell, the fertilized egg or zygote. The first cell divisions of a fertilized sea urchin egg are illustrated in Fig. 5.1. Also, adult body cells (somatic cells) divide, especially during growth of the individuals. But even after growth of the organism is concluded, all our tissue cells are turned over (renewed) many times during life. This constant renewal requires stem cells.

When a cell divides, its genetic information as encoded in the DNA sequence must be evenly distributed between the two daughter cells. This requires faithful replication prior to cell division to obtain two copies of the DNA existing in the mother cell. In eukaryotic cells, during the process of the actual division or mitosis, the DNA condenses into very dense "DNA parcels" that are visible as typically shaped chromosomes by light microscopy and that can then easily be divided between the new daughter cells (see Fig. 2.3). Thus, typical chromosome structures are only formed during mitosis. After replication, two strands of each DNA string are present. These condense into two chromatids with identical DNA sequences and are held together by special proteins, the cohesins, until

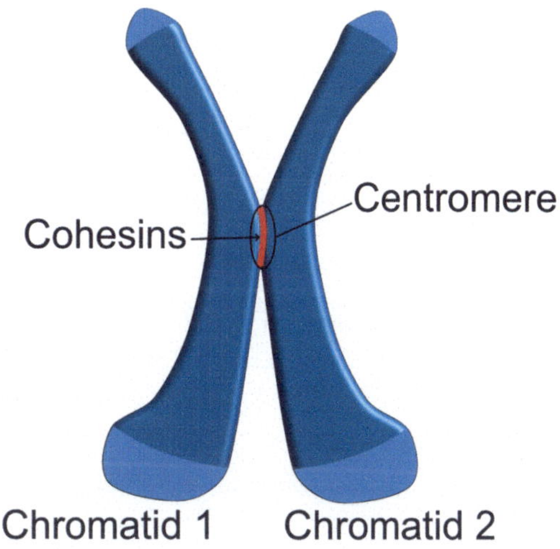

Fig. 5.2 Schematic representation of a chromosome after replication (4n) with its centromere where two chromatids held together by cohesins (red)

mitosis proceeds to the actual separation phase (Fig. 5.2). The place where the chromatids stay connected is called centromere because in human chromosomes, it is localized roughly in the middle of each chromatid. During mitosis, a protein complex (kinetochore) is formed at the centromere. At the kinetochore, microtubules attach and help separate the chromatids during the final steps of mitosis.

The ends of each chromatid are protected by telomere caps. This is necessary to prevent the DNA repair system from mistaking them for DNA breaks. The telomeric DNA contains up to 1000 tandem repeats of the sequence GGGTTA. This repetition is useful because each round of replication shortens the DNA ends on the lagging strand, where a new RNA primer cannot attach anymore. Dividing cells activate an enzyme, called telomerase, which is able to replenish those lost endings. Telomerase activity is especially high in stem cells and in embryonic cells. Somatic cells, in contrast, cannot replenish their telomeric repeats so efficiently anymore, which leads to a "replicative cell senescence" and protects the organism from uncontrolled cell proliferation. However, tumor cells are able to reactivate telomerase, thereby overwriting this proliferation control system.

In order to coordinate DNA replication and cell division, cells pass through a cell cycle (Fig. 5.3). Much of what we know today about cell cycle control had been originally discovered by studying cell cycle-deficient mutants in yeast (Hartwell et al., 1974; Nurse et al., 1976). In constantly dividing cells, replication and mitosis take turns. DNA synthesis occurs during the S phase of the cell cycle. This is followed by a gap phase *(G2)* where replication is checked and mitosis is prepared. Mitosis then allows separation of chromatids, establishing of two daughter nuclei and cytokinesis, the division of the mother cell into two daughter cells. Another gap phase *(G1)* precedes the next replication phase, where cells grow and check the integrity of their DNA to avoid replication of damaged DNA. Moreover, they already prepare the replication origins for the next S phase. During the cell cycle, the cellular amount of DNA changes. Immediately after mitosis,

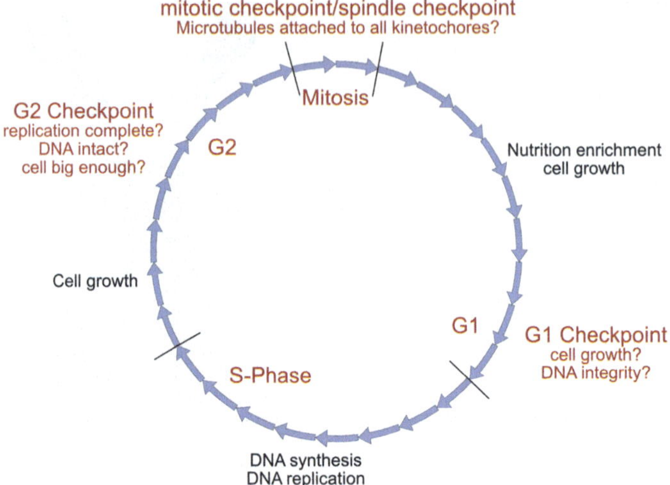

Fig. 5.3 Schematic representation of a cell cycle with the phases G1, S, G2, and mitosis. The checkpoints G1 are also G2, indicated

each daughter cell has one chromatid of each chromosome. The DNA content is 2n (n referring to number of chromosomes) (Fig. 5.4). A diploid cell has two copies of each chromosome. After mitosis, this remains so until the DNA is replicated during the S phase of the cell cycle. While replication proceeds, the DNA content grows continuously until all the DNA is replicated and each chromosome consists of two chromatids. At the end of S phase, the DNA content reaches 4n (Fig. 5.4). After replication, the DNA content remains at 4n until it is divided to 2n during the next mitosis. By following the DNA content of a cell, one can therefore see in which phase of the cell cycle it is. In a single-cell experiment, the DNA content of a cell can be measured by using fluorescent dyes that intercalate with DNA and deliver a DNA concentration-dependent fluorescence signal. In a bulk experiment, the cell cycle of a certain population of cells can be followed with a "*f*luorescence-*a*ctivated *c*ell *s*orter," *FACS*. Here, the DNA of the cells is also labeled with a fluorescent dye. The cells are then suspended and allowed to flow through a measuring point one by one. At this point, the presence of a cell is measured by light scattering and the fluorescence of this cell, which corresponds to the DNA content, is measured after excitation with a laser. As a result, one obtains a diagram where the number of cells with a certain level of fluorescence is displayed (Fig. 5.4). In addition, it is possible to sort the cells according to their fluorescence into different vessels after they have left the measuring point. This allows closer biochemical analysis of a bulk of cells that are in the same phase of the cell cycle.

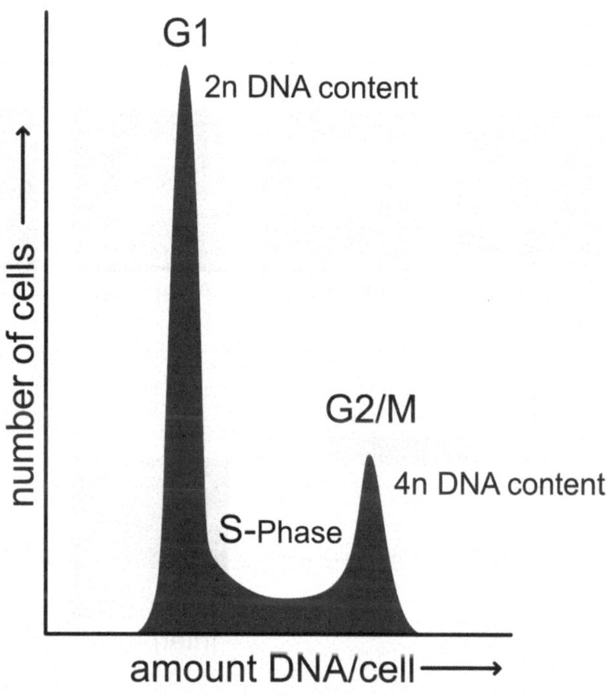

Fig. 5.4 Output of fluorescence-activated cell sorter (FACS) flow cytometry, the number of cells is given on ordinate (*Y*-axis), cells are counted when passing through light scatter detector, DNA content per cell is given on abscissa (*X*-axis), and fluorescence of DNA-binding fluorescent dye is measured when cells pass through fluorescence detector; DNA content equals 2n in G1, 4n in G2, and everything in between in S phase

5.2 Mitosis and Meiosis

Mitosis begins with a "prophase" where chromosomes become visible under the micro-scope. At the same time, centrosomes are arranged at two poles of the cell and the spindle apparatus begins to form. The nuclear membrane dissolves into vesicles and the nuclear lamina disassembles. These processes characterize "prometaphase". Chromosomes are then arranged in the so-called "metaphase plate" in the middle of the cell during "metaphase". The spindle apparatus is fully formed and microtubules are attached to every chromatid. Only then chromatids are moved toward the spindle poles in "anaphase". A new nuclear membrane is built around each daughter nucleus. In "telophase", a contractile ring consisting of actin–myosin fibers is formed in the middle of the cell. It performs contractions, which lead to "cytokinesis" (separation of daughter cells). Two daughter cells are produced, each with identical DNA (Fig. 5.5a). Mitosis occurs in plant cells in a very similar way. However, during cytokinesis, a new cell wall is inserted between the two daughter cells. For this purpose, Golgi vesicles providing the material for the new cell wall arrange at the division plane. They fuse and the cell wall grows from the outside toward the center of the cell until it completely separates the daughter cells (Fig. 5.5b).

Mitosis of somatic cells reproduces two diploid cells from one diploid parent cell. Germ cells, in contrast, only carry one set of chromosomes, and are therefore called haploid. This is necessary because male and female germ cells fuse during fertilization and produce a diploid zygote. Germ cells arise by meiosis. Here, diploid parent cells

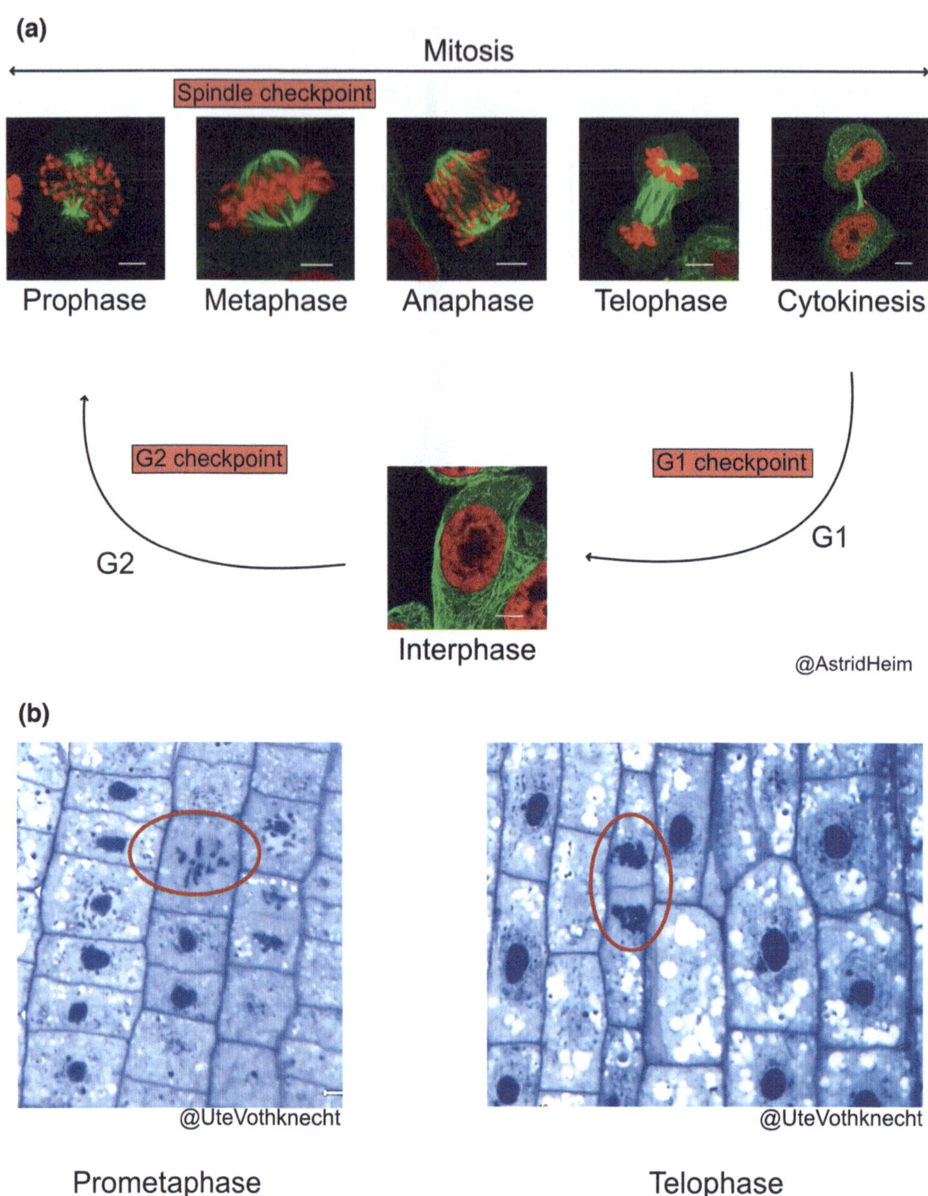

Fig. 5.5 Cell division in animal and plant cells (**a**) Fluorescence microscopic images of HeLa cells taken in all phases of mitosis and in interphase as indicated; DNA is shown by red fluorescence; tubulin is shown by green fluorescence; Chromosomes are visible during mitosis, but not in interphase; (**b**) Images of dividing root cells of *Zea Mays* plant. Prometaphase and telophase are indicated by red circle, staining as in Fig. 1.2

Box 5.1 Chromosomal Aneuploidies

Mistakes in meiosis can result in germ cells with surplus chromosomes or missing chromosomes. After fertilization, such germ cells lead to zygotic aneuploidies, including trisomy (surplus chromosomes) or monosomy (only one chromosome). In humans, only trisomies of chromosomes 13, 18, 21, and X- or Y-chromosomes enable development of the embryo until birth. *Trisomy 13*—children show severe defects in brain, face, eye, and heart anatomy. *Trisomy 18*—children are usually quite small and they have developmental defects which cause many problems with inner organs. Both have relatively low life expectancies and most die as infants. *Trisomy 21* causes the well-known *Down syndrome*. It is much less severe than trisomies 13 and 18. Patients have mental disabilities and often develop gastrointestinal, heart, and other problems. The severity of mental and physical disabilities differs enormously between individuals with Down syndrome. *Klinefelder* syndrome is seen in boys with an additional X chromosome. They do not show very strong symptoms in childhood, sometimes they have learning disabilities. Sexual development may be affected and can result in smaller testes and sometimes increased growth of breasts. An additional X chromosome (trisomy X) in girls only rarely shows physiological or mental abnormalities.

(germ cell precursors) produce haploid germ cells. Mistakes in meiosis may lead to chromosomal aneuploidies (an abnormal number of chromosomes in a cell), (see Box 5.1).

Meiosis consists of two parts, meiosis 1 and meiosis 2. These are preceded by a meiotic S-phase, where all chromosomes are replicated and therefore consist of two identical chromatids (these cells are 4n). During meiosis 1, after arrangement at the metaphase plate, homologous chromosomes are separated and pulled to the spindle poles. Each chromosome still has two chromatids. Meiosis 1 is followed by meiosis 2, which works like mitosis, separating chromatids. Therefore, during the two phases of meiosis, 4 haploid germ cells are produced from one germ cell precursor. This mixes the genetic material from both parents. Meiosis 1 prophase includes an additional "mixing" step when the chromatids of homologous chromosomes get together into close proximity. Now chromatids of the homologous pair cross over and form a chiasma. They break at certain positions and then the chromatids of the different chromosomes fuse again. Thus, paternal and maternal DNA is now already mixed within one chromosome (Fig. 5.6).

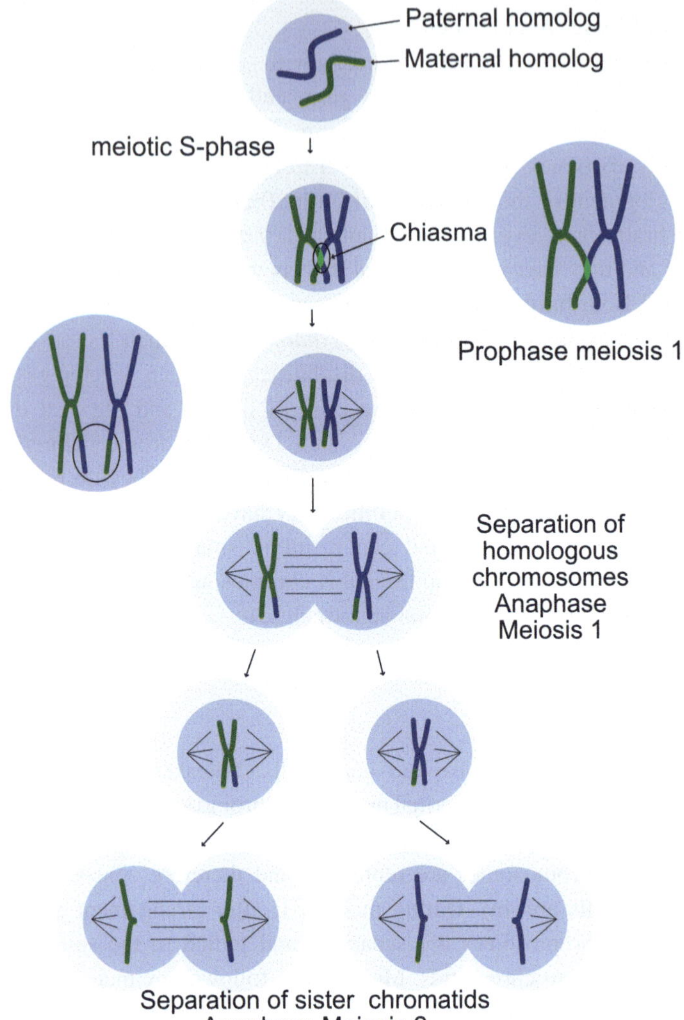

Fig. 5.6 Schematic representation of meiosis: One chromosome pair is shown only; paternal homolog in blue, maternal homolog in green; after meiotic S phase, both chromosomes consist of 2 chromatids, chiasma in meiotic prophase 1 as indicated, in anaphase of meiosis 1, homologous chromosomes are separated and distributed to daughter cells, and these undergo mitosis 2, where sister chromatids are separated in anaphase as in mitosis

5.3 Regulation of Cell Division

Whether or not a cell divides in an adult animal depends on its position within the organism. Is it a stem cell and has to provide new cells to replenish the organ? Is it a terminally differentiated cell of the gut epithelium, which will be lost after it has exhausted its function to digest the food? Is it a neuron of the brain, which has found its stable synaptic connections and will not be replaced during the lifetime of the individual? Is it a bone cell in a growing individual? The information that the cell needs to decide whether to divide or not to divide is received from other cells either in the surrounding or in the blood and tissue fluid of the organism. Such instructive signals are transduced by growth factors binding membrane receptors or by signals from the extracellular matrix. Cells receiving growth factor signals when they are in the G1-phase of the cell cycle may enter a new cycle and divide.

The correct progression through the cycle is ensured by three cell cycle checkpoints (Echalier et al., 2010). With the G1 checkpoint, the cell decides whether it should enter the cell cycle. This depends on the presence of growth factors, which tell the cell that it is in the correct context for division. It also prevents cell division if DNA is damaged. In this case, the transcriptional activator protein p53 accumulates and induces transcription of genes encoding cell cycle inhibitors or cell death activators (see later). The G2/M checkpoint controls completeness of DNA replication and cell growth. A third checkpoint is the spindle checkpoint or metaphase checkpoint. It controls whether all chromatids are targeted by microtubules. This is important to ensure that the two chromatids of each chromosome are moved to opposite poles during anaphase (see Figs. 5.3 and 5.5a).

The molecular agents to lead the cell through these checkpoints are cyclin-dependent kinases (cdks). As the name suggests, cyclins are small proteins the concentration of which changes with cell cycle progression. Cyclins regulate the activity of cyclin-dependent kinases (Fig. 5.7a). In complex with the appropriate cyclin, such cdks phosphorylate target proteins, which then induce the crucial steps of the cell cycle. Without a cyclin bound to them, cdks are not active. Different cyclins are specifically expressed during the individual cell cycle phases (Fig. 5.7b).

G1 cyclins are made in G1 and reach their peak after the G1 checkpoint (in yeast, this is called "start") has been passed. Their transcription is induced by growth factor signaling. Without growth factors, cells will not enter into the cell cycle. G1 cyclins are degraded at the end of G1 to give way to S phase cyclins. S-phase cyclins are beginning to accumulate at the end of G1 and reach their peak concentration at the beginning of S-phase. Activation of their transcription is prevented by the tumor suppressor "retinoblastoma protein", (Rb), (reviewed in Dick and Rubin (2013) until the G1 checkpoint is passed. Checkpoint passage leads to G1 cdks activation causing Rb phosphorylation. This releases transcription factors for S-phase cyclins from inhibition by Rb, initiating S-phase cyclin transcription. S-phase cyclins are degraded in mitotic prophase. During S-phase and in G2, mitotic (M-phase) cyclins are synthesized. They reach their highest level at the beginning of mitosis and are abruptly degraded after the spindle checkpoint is passed and cells start anaphase of mitosis (Fig. 5.7b).

The activity of cyclin-dependent kinases has to be under strict surveillance. As shown in Fig. 5.7b, the G1 cyclin (cyclin D) accumulates in G1-phase of the cell cycle, G1cyclin/cdk is only fully activated just before the beginning of S-phase, and is quickly degraded when S-phase begins. Similarly, the mitotic cyclin (cyclin B) accumulates during the G2-phase, and it only starts to phosphorylate its target proteins when mitosis begins.

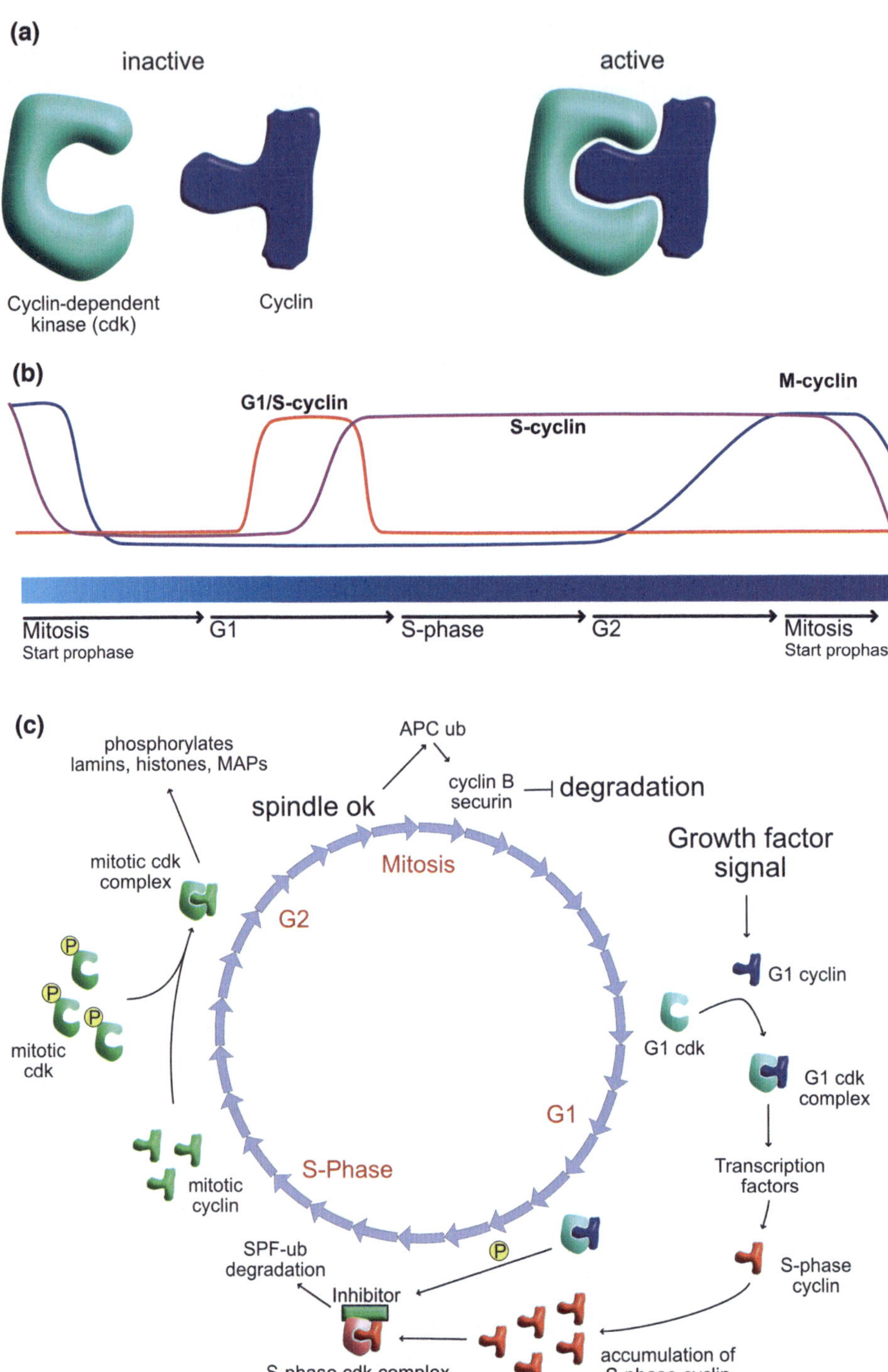
(a)
inactive
active
Cyclin-dependent kinase (cdk)
Cyclin
(b)
G1/S-cyclin
S-cyclin
M-cyclin
Mitosis
Start prophase
G1
S-phase
G2
Mitosis
Start prophase
(c)
APC ub
phosphorylates lamins, histones, MAPs
spindle ok
cyclin B
securin
degradation
Growth factor signal
Mitosis
mitotic cdk complex
G2
G1 cyclin
mitotic cdk
G1 cdk
G1 cdk complex
G1
S-Phase
mitotic cyclin
Transcription factors
SPF-ub degradation
Inhibitor
S-phase cyclin
S-phase cdk complex
accumulation of S-phase cyclin

Therefore, cyclin binding to cdks is not sufficient for initiating kinase activity. For this, in the case of the mitotic cyclin B/cdk1, for instance, an additional phosphorylation of the kinase is required as well as removal of inhibitory phosphates by a phosphatase. These modifications occur when the G2/M checkpoint "clears" the entry of the cell into mitosis. At this point, cyclin B/cdk phosphorylates lamins, allowing nuclear lamina and nuclear envelope disassembly, it phosphorylates histones to complete chromosome condensation, and it phosphorylates microtubule-associated proteins to initiate spindle formation. With all these processes going on, mitosis is under way.

Another way to keep cyclin-bound cdks in check is the activation of cdk inhibitors. These bind to the cdk and efficiently block the active site, preventing the phosphorylation of target proteins. These inhibitors are removed at entry of the cell into the respective cell cycle phase. For example, G1 cyclin/cdks phosphorylate an S-phase cyclin/cdk inhibitor right before S-phase entry. This phosphorylation provides an interaction site for the ubiquitin-ligase S-Phase Factor (SPF), which targets the cyclin inhibitor for ubiquitinylation and proteasomal degradation, thereby ending inhibition of the S-phase cyclin/cdks, which in turn phosphorylate replication factors to initiate S-phase replication of DNA. Cyclin/cdk inhibitors also play a role in cell cycle arrest after DNA damage. When the cell senses DNA damage in the G1-phase of the cell cycle, transcription of a G1cyclin/cdk inhibitor is initiated to prevent cell cycle progression until the DNA is repaired.

At the end of a specific cell cycle phase, the cyclin/cdks are inactivated. An example for this is the mitotic checkpoint (also metaphase–anaphase checkpoint or spindle checkpoint). When the cell senses that all chromatids are safely connected to microtubules, the ubiquitin ligase "anaphase promoting complex" (APC) is activated. It ubiquitinates the mitotic cyclin B and thus submits it to proteasomal degradation. Cyclin B/cdk1-dependent phosphorylation of histones, lamins, and microtubule-associated proteins ceases and anaphase is initiated. In addition, APC activates the protease (named

Fig. 5.7 Cyclin-dependent kinases: (**a**) Schematic representation of kinase (cyan) and cyclin (blue); Kinase is only active with cyclin bound to it; (**b**) Cyclin content of cell during cell cycle, G1/S-cyclin (red line) is produced in G1 and disappears at the beginning of S phase, S phase cyclin (purple line) is produced after G1/S checkpoint has been passed and degraded after passage of G2/M checkpoint, mitotic cyclin (cyclin B, blue line) is produced in G2, and degraded when the spindle checkpoint is passed. Note the slow increase slopes of cyclins and the abrupt degradation slopes; (**c**) Schematic of cell cycle with schematic representation of major molecular mechanisms guiding the cell through the cell cycle; in G1, the cell receives signals from growth factors to produce G1 cyclins activating G1-cdk, G1-cdk phosphorylates transcription factors to activate transcription of S phase cyclins, S phase cyclin-cdk complexes are inhibited by cdk inhibitors before onset of S-phase, inhibitor is abruptly degraded by SPF ubiquitin ligase to allow onset of S-phase, in G2, mitotic cdk remains inactive by inactivating phosphorylations, these are removed cdc25 phosphatase when G2/M checkpoint is passed, active mitotic cyclin-cdk phosphorylates lamins, histones, and MAPs (microtubule-assembling proteins) to induce mitotic prophase with chromatin condensation into chromosomes, nuclear envelope disassembly, and spindle formation, and when spindle checkpoint is passed, APC degrades mitotic cyclin and securin, anaphase is induced, and securin releases separase to separate chromatids "separase") that cleaves cohesins for chromatid separation by ubiquitinylating an inhibitor of this protease (named "securin") and submitting it to degradation by the proteasome.

These examples illustrate that cell cycle progression is tightly regulated. Regulation involves cyclin-dependent kinases, which are themselves subject to stringent control by phosphorylation/dephosphorylation, cdk inhibitors, and ubiquitin ligases.

In the following, the processes that govern one round of the cell cycle, as we know them to date, are summarized and illustrated in Fig. 5.7c.

At the beginning of the G1-phase of the cell cycle, DNA replication complexes assemble at replication origins. When the cells receive growth factor signals, G1-cyclins are synthesized and G1cyclin/cdk complexes are activated. They phosphorylate protein targets and activate distinct transcription factors, which activate genes encoding S-phase cyclins. As a result, S-phase cyclins accumulate, but the S-phase cyclin/cdk complexes are blocked by an inhibitor. Just before S-phase begins, this inhibitor is phosphorylated by the G1cyclin/cdk complex, making it amenable for binding the ubiquitin ligase SPF (S-phase factor), which leads to its transfer to the proteasome and degradation. Now active S-phase cyclin/cdks phosphorylate the preassembled replication complexes inducing DNA replication. Each chromosome is replicated and at the end of S-phase, it consists of two identical sister chromatids. Sister chromatids are held together by cohesin molecules. Already during S-phase and in the G2-phase of the cell cycle, mitotic cyclins are transcribed and translated. Until replication is complete, the activity of the forming mitotic cyclin B/cdks is inhibited by phosphorylation. After passing through the G2-M checkpoint, inhibitory phosphates are removed and mitotic CDK complexes are activated. They target several substrates that are active during mitosis. These include nuclear lamins, histones, and microtubule-associated proteins (MAPs). Lamin phosphorylation induces disassembly of the nuclear lamina and at the same time, the nuclear envelope dissolves into small vesicles. Phosphorylation of histones aids chromosome condensation into distinct microscopically visible chromosomes. Phosphorylation of MAPs is required for the formation of the spindle apparatus. With mitosis progressing, the chromosomes are arranged in the center of the cell forming the metaphase plate. For the mitotic checkpoint to be passed, all chromatids must be attached to microtubules. Then the anaphase complex APC is activated. APC targets securin, an inhibitor of separase, the protease degrading cohesins, for degradation, thus allowing proteolytic cleavage of cohesins, segregation of chromatids, and their moving to opposite poles of the cells. The mitotic cyclin B is also transferred to the proteasome by the APC. This induces the final steps of mitosis where the nuclear envelop and the lamina are re-established, the spindle apparatus falls apart, and cytokinesis is completed. The two daughter cells emerge.

5.4 Regulation of Cell Growth and Proliferation

Cell divisions cause an increase in cell numbers, meaning that cells "proliferate." Proliferation is strictly regulated by factors from the cellular environment. This ensures that cells only proliferate when the conditions for them are right. In the context of a multicellular organism, this is very important to regulate cellular homeostasis, an equilibrium of all cells in the body.

The human body develops from one single cell, the fertilized egg. This cell then divides to make more and more cells. All cells take different fates in order to develop the embryo and later, the adult. Eventually cells specialize to produce every tissue and organ of the body. Cells therefore adapt different shapes and contents to perform specialized functions. This process is termed "differentiation".

According to recent new estimates, the body of a 70-kg adult male (a so-called reference man, 20–30 years old, 70 kg, 170 cm tall) (Sender et al., 2016) contains 30×10^{23} cells. These include 56 different cell types, whereby with a count of around 2.5×10^{13} cells, 84% of all body cells are erythrocytes, followed by 4.9% of platelets, 2.5% of bone marrow cells, and 2.1% of endothelial cells (the lining of our blood vessels). The residual 52 cell types then make 7.5% of the total cell numbers. With only 3×10^{12} nonblood cell types, the human body cell arsenal essentially consists of 90% of blood cells. However, if we would measure the cell masses of the different cell types, muscle cells would make the top by providing 43% of total body cell mass, followed by fat cells with 28%. These cells are very big, and thus, skeletal muscle cells, which form muscle fibers, only make ca. 0.001% of the total cell count, adipocytes make 0.2% (see Table 5.1) for comparison of cell numbers according to recent estimates made by Sender and colleagues (Sender et al., 2016).

Table 5.1 Cell types and their contribution to the total cell number in the human body

Cell type	Percent of total body cells According to Sender et al. (2016)	Contribution to total cell mass of 46 kg for 70 kg "reference male," according to Sender et al. (2016)
Erythrocytes (blood)	84	2.5 kg
Platelets (blood)	4.9	
Bone marrow cells (blood)	2.5	
Vascular endothelial cells	2.1	
Other	2	
Lymphocytes (blood)	1.5	
Hepatocytes (liver)	0.8	
Neurons and glia (brain)	0.6 (0.3 + 0.3)	
Epidermal cells	0.5	
Bronchial endothelial cells (lung)	0.5	
Respiratory interstitial cells (lung)	0.5	
Adipocytes (fat cells)	0.2	13 kg
Dermal fibroblasts (skin)	0.1	
Muscle cells (skeletal muscle)	0.001	20 kg

To keep the cellular homeostasis during embryonic development, growth, and adult life of an animal cell, production (proliferation) and cell differentiation have to be very precisely regulated. In a developing embryo and in growing adults, new cells are produced constantly to allow growth. Moreover, during all adult life, tissue and organs are always renewed. Therefore, our body contains specific adult stem cells that divide when necessary to produce new precursor cells for the replacement of lost or dysfunctional cells. The cellular turnover rates have recently been quantified by Sender et al. (Sender & Milo, 2021). They estimated the lifespan for most cell types based on their death rates and related this to their cellular mass contributions in the aforementioned "reference man." A cell of the epithelium lining the gut only lives for 3–5 days, then, it has to be replaced. In contrast, heart muscle cells, skeletal muscle cells, and glial cells live for years. Only a very small proportion of heart cells is replaced during the lifetime of a human. Blood cells have an average life span of 5–8 days for monocytes and neutrophils, 100 days for erythrocytes, and 500 days for mature T cells. Within these time frames, new cells are produced to maintain the cellular homeostasis in our body. Sender and coworkers calculated that 3.3×10^{11} cells are turned over per day. Most of those are erythrocytes with 65%, followed by neutrophils with 18% and gastrointestinal epithelial cells with 12%. In this way, our blood is basically completely replaced every so often.

These cellular proliferation and differentiation activities have to be controlled very tightly. At older age, such regulation can go awry and this may lead to proliferative illnesses, such as cancer and others. In the following, some basic molecular mechanisms for regulating cell division and growth are explained. They essentially have an impact on the regulation of the cell cycle.

As the cell division cycle only starts when the G1 checkpoint is passed, it depends on the transcription of genes encoding G1 cyclins (e.g., cyclin D). Transcription of such genes is controlled by growth factors.

5.5 Growth Factor Signaling

Growth factors activate receptors at the cell surface. These receptors "transduce" growth factor signals to the cell nucleus where transcription is activated (Fig. 5.8). Many growth factor receptors in animals are receptor tyrosine kinases. "Growth factor" is a rather general term. It also includes extracellular signaling molecules regulating cell survival and death, cell differentiation, and cell growth. Therefore, growth factors are important for all development (Fig. 5.8).

Examples for growth factors are epidermal growth factors (EGF), fibroblast growth factor (FGF), platelet-derived growth factor (PDGF), and nerve growth factor (NGF). All these growth factors are relatively small secreted proteins. They often have many cysteine residues in the primary structures of their extracellular domains and form disulfide bridges, giving them a very distinct tertiary structure. They fit closely onto specific receptors, which are named accordingly, including epidermal growth factor receptor (EGFR),

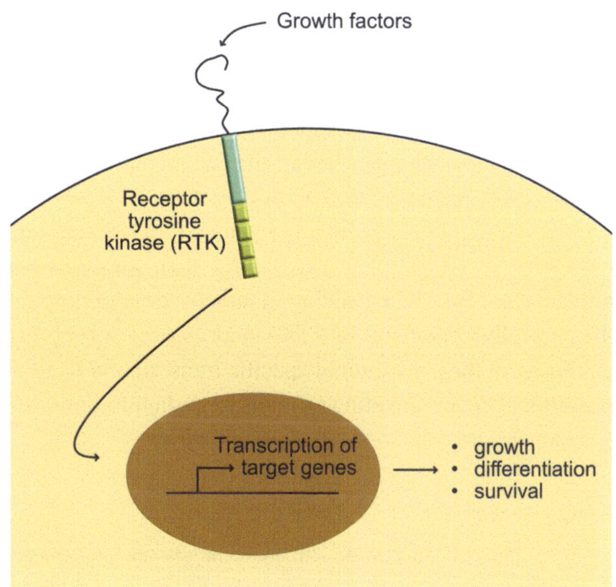

Fig. 5.8 Receptor tyrosine kinase (RTK) signaling: Growth factor binds to RTK on cell membrane, and signal transduction leads to transcription of target genes in the nucleus to induce responses, such as cell growth, cell differentiation, and cell survival

platelet-derived growth factor receptor (PDGFR), and nerve growth factor receptor (NGFR).

The research history of growth factors and their receptors is closely linked with tumor research. It started in 1982 when retroviral proteins were discovered that had the ability to induce cells to proliferate and "transform" them to become tumor cells (Sefton et al., 1982). In this context, the term "transformation" refers to exactly this change of cells from normal to tumor cells. One example is the *Rous sarcoma* virus protein v-SRC. It turned out that this transforming protein was a tyrosine kinase, meaning an enzyme that phosphorylated tyrosine residues of target proteins (Hunter & Sefton, 1980; Sefton et al., 1980). Soon it was discovered that EGFR, PDGFR, and also the insulin receptor had tyrosine kinase activity. In 2001, 90 human tyrosine kinases had been discovered, and half of those were found to play a role in cancer. Gleevec™ was the first tyrosine kinase inhibitor approved by the *FDA* (US *F*ood and *D*rug *A*dministration) as an anticancer drug (reviewed by Hunter and Eckhart (2004)).

As discussed previously, proteins are often phosphorylated in our cells. This provides a regulatory switch because it can change the protein structure distinctly and fast. Most protein phosphorylation events occur on serine and threonine residues. Tyrosine phosphorylation is rare in comparison. However, in tumor cells, the amount of phosphorylated tyrosine residues increases, indicating the importance of tyrosine kinases for cell proliferation.

As mentioned, growth factor receptors are basically receptor tyrosine kinases (RTK). They are transmembrane proteins. Their extracellular part has a very distinct tertiary structure allowing a close fit for the respective growth factors. The intracellular domain has a tyrosine kinase enzymatic activity. In addition, RTKs have many tyrosine residues which are phosphorylated in the process of growth factor signaling. When the growth factor binds the receptor, the tyrosine kinase activity of its intracellular domain is activated. Two receptor molecules form a dimer, whereby the two monomers phosphorylate each other at tyrosine residues These phosphorylated intracellular domains then recruit further signaling molecules to the inner side of the cell membrane including adaptor proteins (Hunter & Sefton, 1980, Sefton et al., 1980), (Fig. 5.9). RTK adaptors strictly only bind to the intracellular domain of the receptor, when certain tyrosine residues are phosphorylated. They have very specific protein regions for their phospho–tyrosine interactions, called SH2 domains (see Box 5.2). Binding of adaptor proteins to phosphorylated tyrosine residues at the intracellular domains of RTKs induces a cascade of further phosphorylation events that culminate in the activation of specific transcription factors in the nucleus and induce transcription of genes encoding proteins important for initiating cell division, e.g., G1 cyclins.

Box 5.2 SRC

Src stands for *Sar*coma. The src gene encodes a nonreceptor tyrosine kinase, which was first discovered as a component of the *Rous sarcoma* virus. When the viral gene was transfected into cells, they lost their growth control and thus obtained properties of cancer cells. Src was therefore classified as one of the "transforming" factors of the *Rous sarcoma* virus. Furthermore, it was published in 1980 by Sefton and Hunter that the SRC protein was able to phosphorylate tyrosine residues. This was the first time that tyrosine residues of a substrate were shown to be phosphorylated by a kinase. Before this, only threonine and serine residues had been identified to be phosphorylated. In the assay, after the phosphorylation reaction, the substrate (in this case, the heavy chain of an immunoglobulin) was degraded into its single amino acids and those were separated by paper electrophoresis. Phosphorylated amino acids were detected due to a γ^{32}P-label of the ATP present in the kinase reaction. In a personal account published in the Cell S216 issue, 2004, Tony Hunter recalls the difficulties to separate phosphorylated tyrosine residues from phosphorylated threonine residues by the methods used at the time. Tyrosine phosphorylation was soon to be shown to be strongly increased in tumor cells. Further cellular proteins including epithelial growth factor (EGF) receptor, insulin receptor, and platelet-derived growth factor (PDGF) receptor were identified as tyrosine kinases. Four domains in SRC were found to be conserved in many proteins. They are now designated SH1, 2, 3, and 4 domains. SH2 is the one, which specifically interacts with phospho-tyrosine, but not with non-phosphorylated tyrosine residues. SH3 interacts with the ras-GEF SOS.

The single components of an exemplary pathway which leads to activation of the cell cycle by inducing cyclin transcription are described in the following. The required transcription factors in this pathway are activated by phosphorylation. The responsive kinase is called "*m*itogen-*a*ctivated *k*inase," *MAP* kinase (MAPK). Mitogen stands for a factor

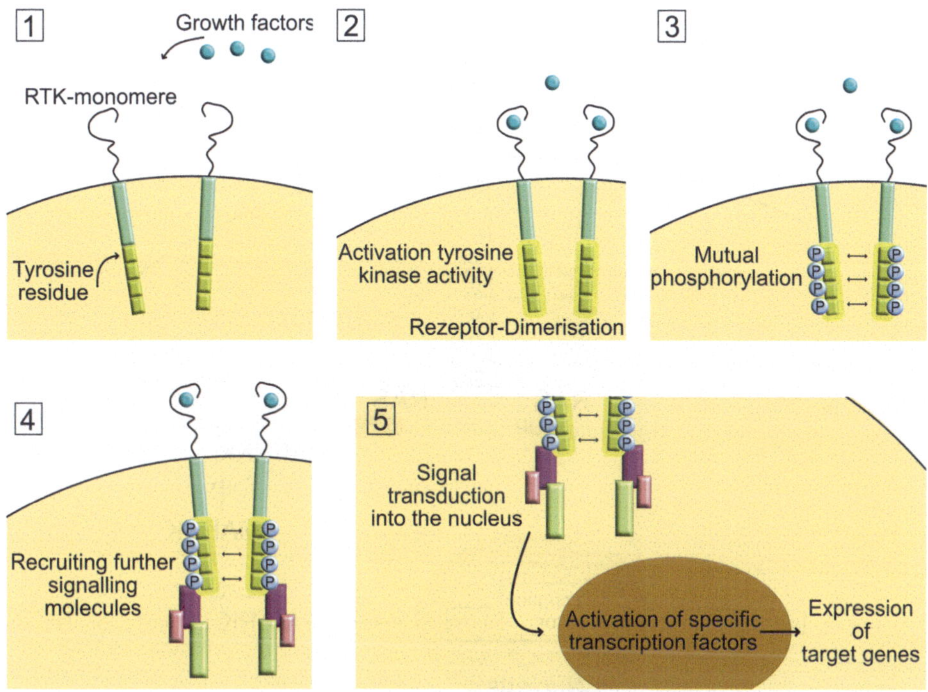

Fig. 5.9 Signal transduction for growth factor signaling at RTKs: (1) Growth factor binds to extracellular domain of receptor tyrosine kinase (RTK); (2), (3) Receptor dimerizes and tyrosine kinase activity is activated; (3) Tyrosine residues at intracellular part of receptor are phosphorylated; (4) Adaptor molecules and signaling factors (purple, pink, and green boxes) are recruited to tyrosine phosphates; (5) Signal is conveyed into nucleus, e.g., by activation of kinases which phosphorylate and thereby activate transcription factors, and then, target genes are expressed

that induces mitosis, such as a growth factor. MAP kinases are controlled by phosphorylation themselves. To be activated, they are phosphorylated by MAP kinase kinases (MAPKKs), which in turn have to be activated by phosphorylation, conducted by MAP kinase kinase kinases (MAPKKKs). MAPKKKs are thus the most upstream part of the cascade. They are the target for the growth factor-bound receptor. A very prominent activator of MAPKKKs is a small GTPase, named ras. Ras has a GTP-bound active state and a GDP-bound inactive state (it is a small GTPase switch-protein, see Box 3.3.). In order for GTP to bind, it requires an exchange factor, GEF. The GEF for ras is called "son of sevenless" (SOS). GTP-bound ras activates MAPKKKs (Fig. 5.10), (Rozakis-Adcock et al., 1993; Chang et al., 1994; Thomas & Wassarman, 1999).

Cells contain several MAP kinases, which are activated by different MAP kinase cascades. The general principle of this activation is the same, but the result is not always proliferation. Specific MAP kinase cascades are also involved in the regulation of differentiation processes and cell survival. MAP kinase activation pathways are more complex than described here. MAPKKKs, for instance, can also be activated by phosphorylation, including that mediated by MAPKKK kinases (MAPKKKKs), but also by others.

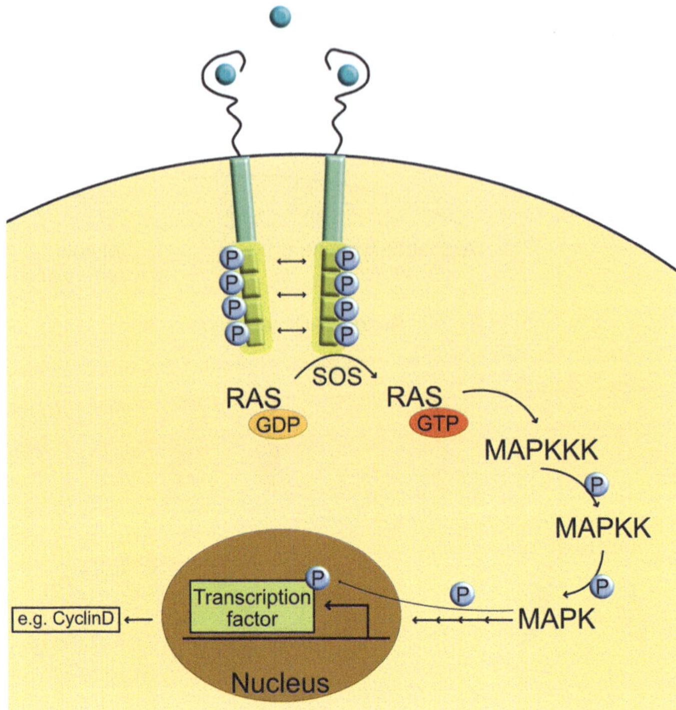

Fig. 5.10 Schematic representation of activation of MAP kinases by RTK signaling: RTK is activated by binding of growth factors, and tyrosine residues at intracellular domain of receptor are phosphorylated; adaptor proteins (not shown) recruit guanyl nucleotide exchange factor (GEF) for small GTPase Ras (SOS) to the membrane, Ras binds GTP and activates MAPKKK, MAPKKK phosphorylates MAPKK, MAPKK phosphorylates MAPK (MAP-kinase), and MAPK translocates into nucleus and activates transcription factors, for example, those specific for G1 cyclin genes, leading to activation of mitosis in G1

Box 5.3 Discovery of Growth-Factor Signal Transduction Pathways

The discovery of the signal transduction chain from growth factor activation by ligand binding to a receptor via activation of ras was a great example for biochemical and genetic research coming together to decipher a conserved and important cellular pathway. While biochemists performed protein–protein interaction studies, *Drosophila* geneticists studied the development of the R7 photoreceptor cell responsible for UV photoreception in the facet eye of the fly. In UV-blind fly mutants, they found a mutated receptor, "sevenless" and a mutated transmembrane ligand present on R7 neighboring cells ("bride of sevenless"). "Sevenless" turned out to be a receptor tyrosine kinase. Loss of function of this receptor could be compensated by gain-of-function mutants in three genes. One encoded the adaptor protein *Grb* possessing

Box 5.3 (continued)

a SH2-domain capable of interacting with phosphorylated tyrosine and an SH3-domain capable of interacting with a ras-GEF (see Box 5.2). The second gene encoded a ras-*GEF*, that was called "son of sevenless, *SOS*," and the third encoded the Drosophila *ras*-gene. All three are conserved in animals where they had been found to interact with each other by biochemical investigations in human cells. This is how the name *SOS* found its way into the nomenclature of this signaling pathway.

Summarizing our example for activation of the cell cycle by growth factor signaling, growth factors bind to receptor tyrosine kinases. These dimerize and each part of the dimer phosphorylates tyrosine residues at the intracellular domain of its partner. Such phospho–tyrosine switches engage adaptor proteins, which recruit the rasGEF SOS, leading to activation of ras. Ras initiates the MAP kinase cascade, activating transcription of cell cycle genes (Fig. 5.10). Ras is a very prominent oncogene. In many tumors, gain-of-function ras mutations have been found, which maintain ras in its activated state and so allow cell cycle progression of the tumor cell in the absence of growth factor signals.

5.6 Apoptosis

To maintain a stable balance between all cells in our body, a close integration of cell proliferation, cell differentiation, and cell death is required. Cell death in this context is a tightly regulated cellular suicide program; in animals, it is called "apoptosis". Programmed cell death is distinguished from "necrosis", an accidental cell death, which happens through injury or lack of blood supply, extreme heat or cold, among others. The main difference between necrosis and apoptosis is that necrosis causes cell contents to penetrate tissues and induce inflammation, whereas apoptosis is almost invisible as cells shrink and finally get phagocytosed by neighbors or macrophages without leaving any traces.

Apoptosis is especially important for developmental processes. Examples are the removal of interdigital tissue during mammalian embryogenesis or the removal of the tadpole tail during metamorphosis. In adults, apoptosis is used to maintain homeostasis between different tissues and to remove damaged cells, especially those with damaged DNA. Morphologically, apoptosis starts with DNA degradation, causing extreme changes in chromatin. During its degradation, DNA condenses into large aggregates. The nuclear lamina is degraded and the cytoplasm shrinks. The plasma membrane remains intact, although larger cells can break into parts; however, each part remains surrounded by its plasma membrane. This breaking up can be observed as "blebbing." The whole process is guided by a molecular program, and hence, apoptosis is also called "programmed cell death" (Fig. 5.11). In the center of initiating this program are proteases, called "caspases".

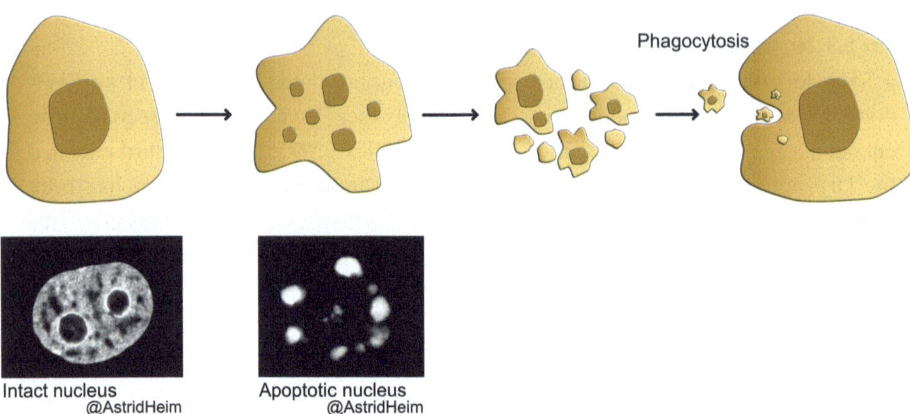

Fig. 5.11 Morphological changes during apoptosis: Cells are shown in yellow with nucleus in light brown, chromatin condensation, break-up of nuclear envelope, nuclear fragmentation, cell blebbing, cell fragmentation, and finally, phagocytosis of nuclear bodies by professional phagocyte or bystander cell; photographs depict HeLa cell in interphase with normal chromatin and HeLa cell in apoptosis with condensed and fragmented nucleus

These are *c*ysteine proteases (with a cysteine residue in their catalytic site), which tend to cleave proteins after *asp*artate residues—both features counting for the name *casp*ase. Caspases represent a large protein family consisting of initiator caspases and executioner caspases. All caspases have prodomains of varying lengths and consist of a large and a small subunit (Fig. 5.12). The active site cysteine residue is present in the large subunit. However, the rest of the active site is distributed between the large and the small subunits. In this way, pro-caspases are inactive. Only after proteolytic removal of the prodomain and cleavage between the large and the small subunits, an active caspase molecule is formed. It constitutes a heterotetramer with two large and two small subunits arranged in such a way that both parts of the corresponding parts of the active sites are able to form a functional catalytic center. This is an important mechanism preventing caspases from cleaving their substrates in cells in the absence of apoptotic stimuli (Fig. 5.12). Initiator caspases have long prodomains and are activated first in response to apoptotic stimuli in an autocatalytic process, which involves concentrating them in large complexes, called "apoptosomes". Apoptosomes can be formed after extracellular death factors bind to death receptors on the cell surface. This is an extrinsic signal. It is activated when cells are no longer needed, e.g., certain immune cells are not needed any more, when the pathogen is defeated. Apoptosomes can also be induced by intrinsic death signals, which might respond to DNA damage. In this case, interestingly, mitochondrial factors are important. Mitochondria are protected from disintegration by proteins of the Bcl-2 family. Bcl-2 itself is an oncogene, which is dramatically overproduced in lymphocytes of B-cell lymphoma patients. This overproduction prevents these lymphoma cells from undergoing programmed cell death. Normally, other proteins of the Bcl-2 family are activated by intrinsic apoptotic signals, such as DNA damage, and interfere with the protective function

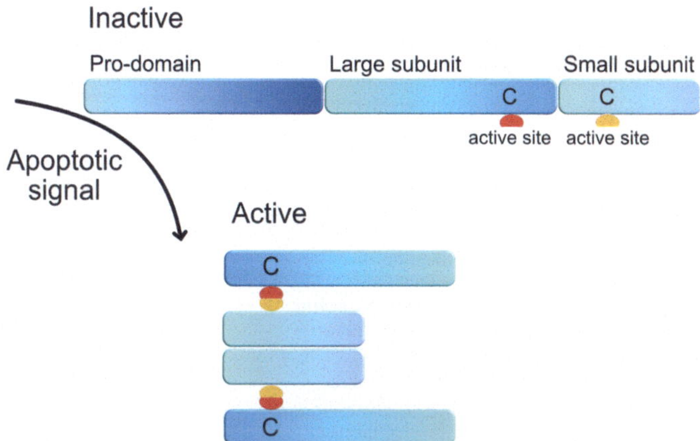

Fig. 5.12 Schematic representation of caspase activation: Caspase prodomain, large subunit, and small subunit are illustrated by blue boxes, and part 1 (red) and part 2 (orange) of active site with cysteine are indicated; apoptotic signal induces proteolytic processing of pro-caspase, prodomain is removed, and large and small subunits are separated, caspase subunits without prodomain assemble as tetramers, large and small subunits of each caspase molecule build the active sites, and the tetramer possesses two active sites, each indicated by a dot comprising a red and an orange part

of Bcl-2. This leads to release of cytochrome C from the mitochondria. Cytochrome C plays an important part in assembling apoptosomes and thereby inducing apoptosis in damaged cells.

The density of caspase molecules in apoptosomes is regarded as inducer of their autocatalytic processing leading to formation of catalytically active heterotetramers. Activated initiator caspases induce a proteolytic cascade where more and more executioner caspases become active. The targets that active caspases degrade are specific. They include inhibitors of DNAases, which after being degraded release active DNAase. These DNAses cleave nuclear DNA at linker regions between nucleosomes, thereby abolishing all functioning chromatin. DNA of apoptotic cells, when analyzed in agarose gel electrophoresis, appears as a ladder with nucleosome-sized DNA fragments (Shi et al., 1990). The DNA fragments then collapse, leading to the typical appearance of apoptotic nuclei as illustrated in Fig. 5.11. The cytoskeleton of apoptotic cells is also destroyed by caspase-mediated degradation processes. A third group of caspase substrates involves signaling molecules, which regulate catabolic, proliferative, or survival processes, opening the path of destruction for the cell. At the end of this way, the apoptotic cell displays death signals on its surface, which allow macrophages to recognize and phagocytose it. Apoptosis, like mitosis, is a very fast process lasting about 1 h (Fig. 5.13). In a normal tissue, at any time, only a few apoptotic cells are detectable. This is the reason why it was so difficult to discover apoptosis. The Scottish researcher Andrew Wyllie was the first to morphologically describe programmed cell death in several tissues and to propose the term "apoptosis" for this phenomenon (Kerr et al., 1972).

Proliferation and apoptosis are two opposite cellular pathways that ensure healthy relationships between all cells in multicellular animals. Accordingly, when they get out of

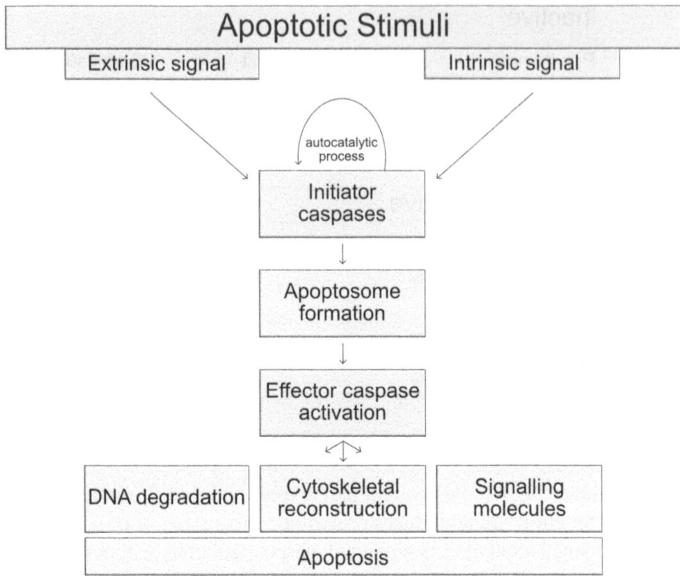

Fig. 5.13 Flowchart showing activation of apoptosis: Extrinsic signals from death receptors or intrinsic signals working through mitochondrial disintegration recruit pro-caspases (initiator caspases) into apoptosome, pro-caspases come into close proximity to each other leading to their activation by proteolytic processing, initiator caspases activate effector caspases, effector caspases degrade very specific molecules including inhibitors of DNAse, leading to degradation of nuclear DNA, components of the cytoskeleton leading to cytoskeletal reconstruction (see cell blebbing), and signaling molecules stopping cell division and differentiation processes as well as metabolic activities

control, the organism can be damaged. An impressive example for a loss of healthy cellular homeostasis is tumor formation. Many mutations in genes encoding proteins involved in regulating proliferation or death pathways are found in tumor cells. The ras gene, as described above, is in many tumors mutated in such a way that it becomes hyperactive. Proliferation is then switched on through MAP kinases even in the absence of growth factors. This leads to uncontrolled proliferation. Ras is therefore called an "oncogene". On the other hand, oncogenes that prevent apoptosis are also often hyperactive in tumor cells. An example is Bcl-2 (B-cell lymphoma). It protects cells from apoptosis, but in B-cell lymphoma cells, the *Bcl-2* gene gets under the control of a very active promoter, which is used for transcribing immunoglobulin genes. Therefore, a lot of Bcl-2 protein is produced and this blocks apoptosis. In contrast, genes that encode proteins inhibiting cell cycle progression, such as the retinoblastoma protein, are inactivated in tumor cells. The same is true for proteins, which induce apoptosis or cell cycle arrest, like p53. These are called "tumor suppressors". Thus, tumor cells have activated proliferation pathways and inactivated apoptotic pathways (Fig. 5.14).

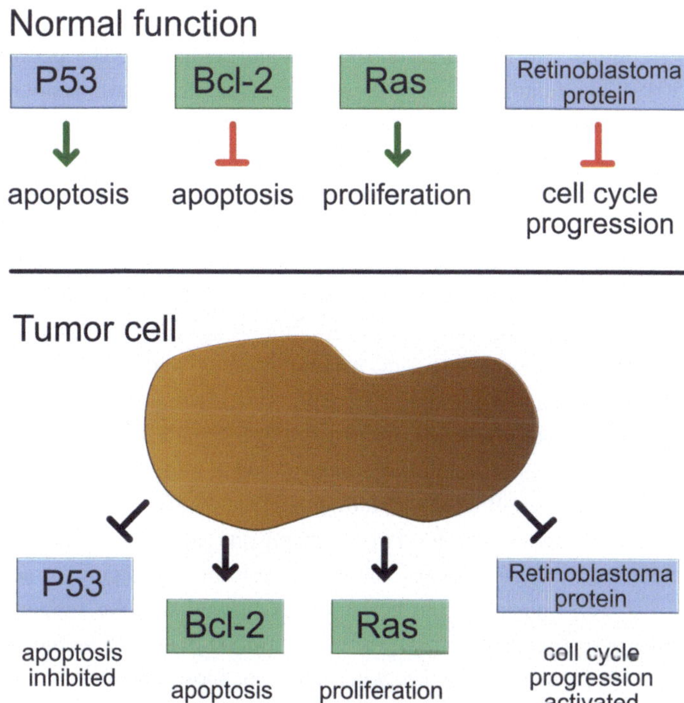

Fig. 5.14 Oncogenes (green) and tumor suppressors (blue): The normal function of a tumor suppressor like p53 and retinoblastoma protein is to either induce apoptosis or to block cell cycle progression (Retinoblastoma) or both (p53, only apoptosis is shown); The normal function of oncogenes is to inhibit apoptosis (Bcl2) or to stimulate cell division and proliferation (Ras). In tumor cells, tumor suppressors are inactivated; therefore, apoptosis is blocked and cell division is promoted; Oncogenes are hyperactivated in tumor cells; therefore, again apoptosis is blocked and cell cycle progression is promoted

5.7 Viruses

According to the Latin–German Dictionary by Georges, a virus is a "*natürliche zähe Feuchtigkeit, Schleim, Saft*" (natural viscous moisture, mucus, juice) of plants and animals but also the "poison of snakes" (Georges, 1910). This may have inspired our current terminology, which characterizes a virus as an infectious agent passing filters with pores <200 nm. We know viruses that cause diseases including childhood diseases, such as polio, measles, small pocks, *Varizella Zoster*, and plant disease like tobacco mosaic disease.

Viruses are composed of proteins and nucleic acids. The nucleic acids usually code for 10 (HIV) to 70 (bacteriophages) genes only. The proteins they encode function as polymerases for nucleic acids, and as structural proteins, which constitute the envelopes of viruses. Viruses cannot replicate without a host cell. This is important. Therefore, it can be debated whether they are truly "alive." Nevertheless, our world is filled with viruses and most of them are not pathogenic! Latest sequencing efforts estimate the number of viruses on earth to 10^{33}, compared to 10^{31} bacteria, 10^{25} stars and prospective 10^{10} people (Moelling & Broecker, 2019). The function of all of these viruses for life on earth is not really understood. In our body, each of us carries 10^{9} virus particles in the gut, throat, and skin and 10^{7} such particles in blood and urine (Popgeorgiev et al., 2013). The nucleic acids of those viruses are different. Some carry DNA, which may be single-stranded or double-stranded, some carry RNA, either double-stranded, single (+) strand RNA or single (-) strand RNA. Some are retroviruses, which reverse transcribe their RNA and integrate into the genome of the host before they replicate. They get help from our cells through long terminal repeat (LTR) DNA. In our genome, protein-coding sequences only make 1.5% of all sequences; 25.9% are intron sequences, which contribute to the immense variability of gene expression. An extreme example of alternative splicing in a single gene is the "Down syndrome adhesion molecule" with 30,016 possible splice patterns; 13.1% of our genome sequences are short interspersed nuclear elements (SINEs) coding small transposons with RNA as intermediate and small RNAs including tRNAs; 20.4% encode long interspersed nuclear elements (LINEs) coding for large transposons and their reverse transcriptase enzymes. In addition, retrotransposons, small transposons and repetitive sequences are encoded. The rest are sequences from heterochromatin and singular sequences with as-yet-unknown function (Fig. 5.15). Transposons are mobile genetic elements, which insert themselves everywhere in the genome, when they get loose. While bacteria only have DNA transposons with recognition sequences and a transposase gene, encoding an enzyme capable of cutting out the transposon and inserting it in a different place (Fig. 5.16a), humans additionally have retrotransposons, which are transcribed into RNA and then

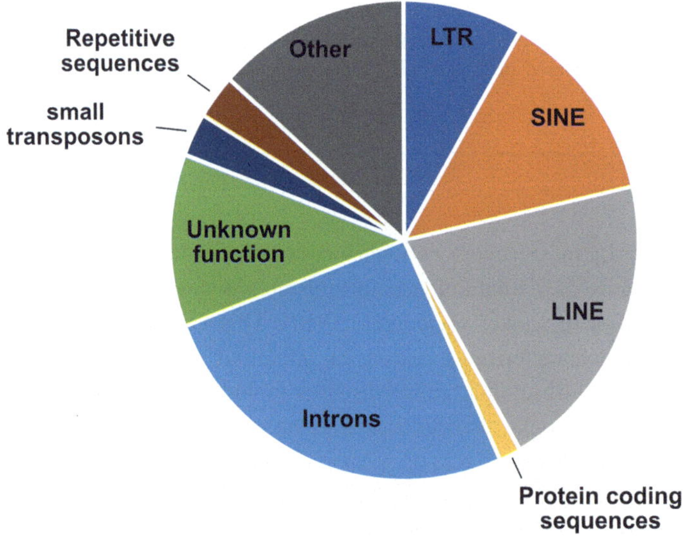

Fig. 5.15 Diagram showing nuclear DNA sequences comprising protein-coding sequences, intron sequences, and others as indicated

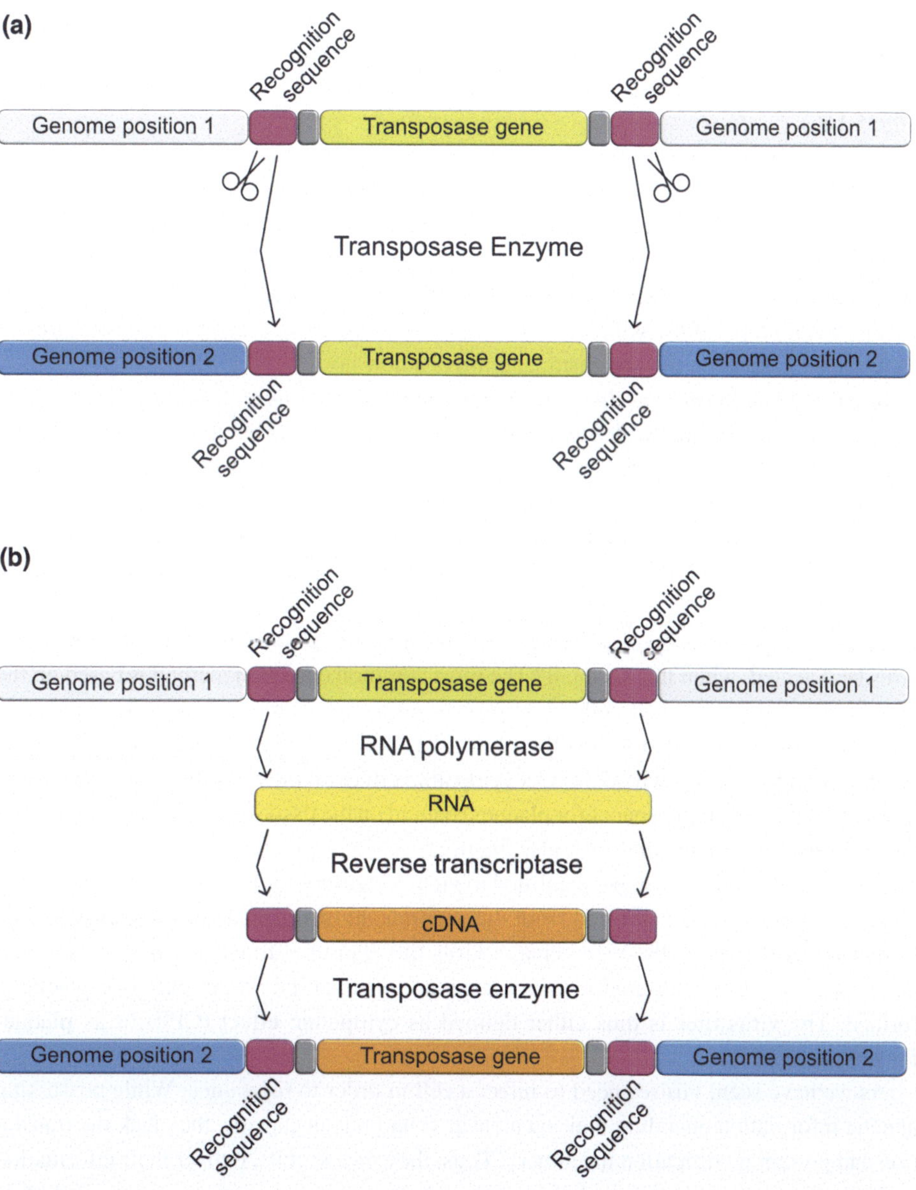

Fig. 5.16 Transposons and retrotransposons: (**a**) Transposon is composed of transposase-encoding sequence (yellow) and recognition sequences (pink). When activated, transposase gene is transcribed and translated. Transposase nuclease activity cuts recognition sequences from genomic position 1 (grey). Transposon is then inserted into new genomic locus (genome position 2 blue); (**b**) Retrotransposon: Recognition sequences (pink) flanking transposase gene (yellow), which also encodes reverse transcriptase; transposase gene is transcribed into RNA by RNA polymerase, translated and RNA is reverse transcribed by reverse transcriptase into cDNA (orange), which can be inserted into new genomic locus (blue) by transposes enzyme

reverse transcribed into cDNA, which is later inserted into the genome at a different place. It should be pointed out here that our genome indeed encodes enzymes capable of reverse transcription, which may also help to insert gene therapeutic RNA into our genomes (Fig. 5.16b).

Thus viruses, in contrast to transposons, are mobile genetic elements, which can leave the cell. Retroviruses, for instance, have this ability. They bring single-stranded RNA into the cell together with a reverse transcriptase, which yields cDNA that is integrated into the genome of the host. After a latent phase, it is possible that such retroviruses are mobilized by unknown triggers and many copies of its RNA are transcribed. These RNAs encode capsid envelope proteins, which allow new virus particles to assemble and the virus to leave the cell and even the organism, and infect a new host.

In 2019, a giant virus was discovered. It is named Megavirus chilensis and has a size of ca. 1 μm, thus not fitting the definition of a virus as based on its passing through a 200 nm filter anymore. Yet, this virus, although it even encodes some ribosomal proteins, needs a host for its replication and is not independently replicating. A similarly big virus is the Pandora virus with 2.5 million base pairs and a size of 1.2 μm. Thus, there are still surprises to be expected in our quest to understand the essence of viruses (Koonin & Yutin, 2018).

Turning back to our "typical" virus particles that do fit into the definition, how can a virus be detected, when it is so small? The most common detection method is based on the property of a virus to infect cells. A virus containing solution, e.g., taken from the nose swap of a patient is added to a cell culture. If it is capable of infecting the cells in the culture, soon its activity will show as a cytopathic effect on the cells. Infected cells sometimes die and leave empty spaces (or plaques) behind on the tissue culture plate. Sometimes, the cells also fuse. They appear under the microscope as giant cells with several nuclei. These cytopathic effects can be quantified to yield a measure for the amount of virus (virus titer). The virus can be harvested from the supernatant of infected cells and enriched by further infecting new cells. Vice versa, a virus titer can be estimated, when the solution containing the virus is diluted in many steps up to the point, where only one plaque is formed. The virus titer is thus either defined as **c**yto**p**athic **e**ffect (CPE) or as **p**laque-**f**orming **u**nit (PFU).

As we have seen, viruses need to infect a cell in order to reproduce. While possessing genetic information encoding their own envelop and nucleocapsids, they lack the translation and protein modification machinery. Thus, they invade cells, deliver their information in the form of a nucleic acid and capture the whole protein synthesis apparatus including ribosomes, t-RNAs, and enzymes delivering post-translational modifications. How this is done is different for every virus. Obviously, it depends on the virus type with respect to the genetic material. But even two RNA viruses, such as SARSCoV2 and Influenza, have very different ways to exploit infected cells for their reproduction. Nevertheless, there are five essential phases of virus infection. These include the entrance of the virus into the cell, and the unpacking inside, translation of viral proteins, virus genome replication, assembling of a new envelope with the replicated nucleic acids, and release of fresh viruses from the cell (Fig. 5.17). In the following, three examples for the infectious cycle of viruses will be

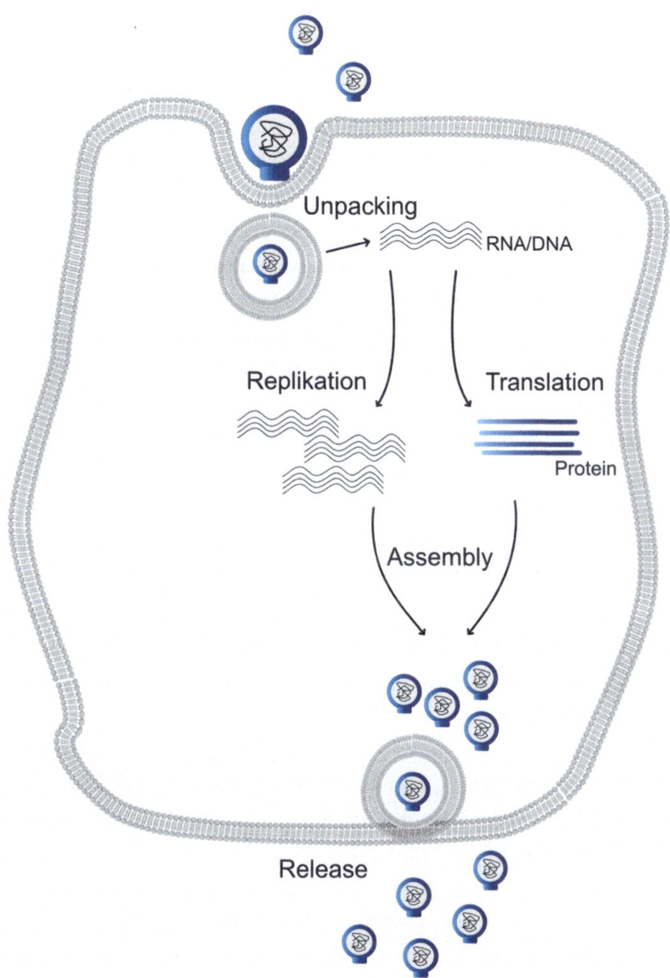

Fig. 5.17 General phases of virus infection: (1) Virus enters cell; (2) Virus is unpacked; (3) Translation of viral proteins; (4) Replication of viral genome; (5) Assembly of new virus particles and release of virus from the cell, nucleic acids, black, proteins, blue

discussed as examples. These include the tobacco mosaic virus in plants, the SARCoV 2, and a retrovirus. Many more examples are discussed in an impressive series of lectures by Prof Vincent Racaniello from Columbia University that are accessible via the following link on Youtube: (https://www.youtube.com/playlist?list=PLGhmZX2NKiNktEViriylQ3haM_a_kAfbk).

The first described virus was the Tobacco mosaic virus (TMV), a plant virus that has been shown to survive even in cigarettes made from infected tobacco plants (Okada, 1999). The Tobacco mosaic virus consists of a 300-nm-long tube, which self-assembles in infected cells from 2100 identical subunits of a small protein with a molecular mass of 18,000. This tube then envelops a 6400 bases long single-stranded RNA molecule.

This virus was discovered in 1882 by Adolf Mayer and also described in 1892 by Dimitri Iwanowski and in 1898 by Martinus Willem Beijernik as "*contagium vivium fluidum*." In 1929, an assay was developed which allowed to follow the tabacco mosaic virus in infected leaves (Necrotic local lesion assay.) In 1935, Bernal and Klug isolated the virus and crystalized it, which secured them the Nobel Prize in 1936. Only later, it became clear that the virus could be reconstituted from an isolated purified protein and RNA fractions, and that the RNA transferred the information for viral production into the leaf. The RNA was later sequenced and contains the gene for the RNA polymerase complex allowing replication of the nucleic acid part of the virus in the cell. In addition, it encodes a "movement protein," which allows the distribution of viral RNA through plasmodesmata to spread the infection, and it encodes the 18,000 kDa capsid protein, which assembles into the tube.

Corona viruses are also RNA viruses. The first one, the infectious bronchitis virus of chickens, was discovered and first described in 1939 by Beach and Schalm. For obtaining large amounts of this virus, the material from infected chicken, after passing it through a Berkefeld filter with a 200 nm pore size was inoculated in the allantois of chicken eggs, where it could be reproduced and harvested in large enough amounts to perform electron microscopy. Cultivation of viruses in chicken eggs is a well-known method to obtain virus for vaccination. While such virus is passed through different eggs, its virulence becomes weaker and weaker until it can be injected in a "weakened" form into patients for immunization. This is the way of vaccine production for many viruses, including influenza, herpes simplex, pox, and mumps.

The infectious bronchitis virus showed rounded protrusions on electron micrographs of purified virus particles on its surface and therefore was named Corona in 1968 by Almeida (Berry & Almeida, 1968). The spike protein, S protein is responsible for these structures. It is integrated in intracellular membranes, which it takes from of host cells (it is therefore an "enveloped" virus in contrast to viruses without membrane envelopes, e.g., adenovirus and TMV), together with the M protein, an integrated membrane protein with three transmembrane domains, and a smaller E protein. Inside this envelope, the nucleocapsid protein (N protein), a basic phosphoprotein with an RNA-binding domain, packages the RNA (Fig. 5.18a). All proteins are encoded by the RNA, which is bound by the N protein. The infectious cycle of the SARSCoV 2 corona virus has now been studied in great detail. The virus attaches to the cell membrane through angiotensin-converting enzyme (ACE), a transmembrane peptidase responsible for processing of proangiotensin to angiotensin, a blood pressure-increasing peptide. This interaction provokes internalization of virus-bound ACE into an endosome. By fusing with the membrane of the endosome, the spike protein leads to uncoating of the RNA and release into the cytoplasm. The RNA is ready to be translated by cellular ribosomes and therefore all viral proteins are produced, including the viral RNA replicase, which produces the (–) strand of the RNA and many new (+) strands. These (+) strands are transported to the ER–Golgi intermediate compartment (ERGIC). The viral envelope proteins are co-translationally localized into the membrane of the ER. They are then assembled into new virus particles in the ERGIC,

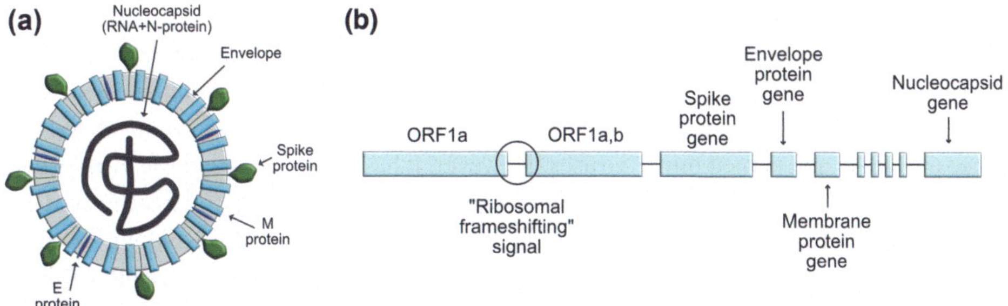

Fig. 5.18 SARS-CoV2-Virus: (**a**) Schematic representation of virus structure showing nucleocapsid with N protein and RNA; Envelope comprising intracellular membranes of the virus host cell where viral small E protein and viral three transmembrane-domain protein (M protein) are inserted, and spike protein assembling on surface of envelope is shown in green; (**b**) Schematic representation of RNA-genome sequence of SARSCoV2 virus: ORF1a and ORF1a, b encode RNA polymerase, ribosomal frame shifting signal creates two ORFs, translating into 2 polyproteins, which are processed into 16 single proteins; Spike protein gene, envelope protein gene, M-protein gene and nucleocapsid gene are indicated

where they meet with the RNA and the nucleocapsid protein after it has also been translated. The virus is further transported within the Golgi, and its surface proteins are glycosylated. Eventually, large Golgi vesicles (Golgi sacks) fuse with the plasma membrane and release the virus. The virus uses the genetic material very economically. All necessary information for producing new virus particles within the host cell is stored in one transcript with ca. 30,000 nucleotides only (Fig. 5.18b). However, a number of regulatory tricks enable the production of diverse proteins. These involve a ribosomal frameshifting process, whereby two different large proteins (pp1a and pp1ab) are translated from the first half of the virus genome, where pp1a encodes the RNA polymerase (replicase). By proteolytic processing of pp1ab, about 16 additional smaller proteins (nsp1–16) are produced, including two proteases, a cell cycle inhibitor to stop cell division of the host cell, and some proteins for anchoring the replicase within the membrane. In addition, from the second half of the virus genome, so-called subgenomic RNAs are produced which enable translation of the spike, M, E, and N proteins (Lai & Cavanagh, 1997; Fung & Liu, 2019).

The spike protein is also known to get to the cell surface alone and to reside in the plasma membrane by itself. This causes fusion of cells and the formation of large multinucleate syncytia, allowing spreading of the virus without production of complete virions, similar to the tobacco mosaic virus, which spreads via plasmodesmata with the help of the movement protein.

Retroviruses have a special enzyme, a reverse transcriptase. Their life cycle involves an additional step. They enter the cell and retrotranscribe their RNA into DNA, which is integrated into the genome of the host cell. At some point, their DNA can be transcribed and new virus particles are formed in the cytoplasm of the cell. The first retrovirus to be discovered was the avian leukemia virus found in 1908 by Ellermang and Bang, as a filterable substance apparently transmitting Erythroleukemia (Ellermann & Bang, 1908). The next one in 1911 was described by Peyton Rous as a filterable extract from chicken sarcomas (Rous, 1967). Because these viruses had been isolated from tumor tissue and retransmission into chickens had promoted formation of tumors, they were considered the cause of tumors and named tumor viruses. However, it has to be said that these viruses can be detected in healthy animals and humans where they also cause antiviral immunity. This has been discussed in detail by Peter Duesberg who had been involved in isolating a mouse leukemia virus in 1966 (Duesberg & Robinson, 1966). He summarizes a wealth of data, which clearly show that such alleged tumor viruses including lymphotropic viruses at least are not sufficient to cause tumor formation (Duesberg, 1987).

Nevertheless, with retroviruses in hand and a cellular assay, which could detect virus transmission in cell cultures (Rubin & Temin, 1958), scientists began to ask which of the molecular components of these viruses was involved in tumor formation. They discovered a number of viral proteins, which were alone able to "transform" cells, which means cells transfected with their genes adopted properties of tumor cells with aberrant cell divisions. Interestingly, these factors included proteins that are also encoded in animal genomes, such as the small GTPase ras, as discussed before a member of an important growth control pathway, the two tyrosine kinases src and abl (see Box 5.2), and the trancription factor myc, which is considered an oncogene. These genes are therefore found in the literature in two versions, the viral version, which is indicated by v as, for instance, in v-ras, and the cellular version, which is pronounced by c, as in c-ras. In summary, this dissection of the molecular components of retroviruses has shed immense light on the nature of human oncogenes and also on viral strategies for replication. Yet, the functions of viruses for biology and human health are not really understood.

Test Yourself

1. How does 1 m of DNA fit into a nucleus of 10 μm diameter?
2. Name the phases of mitosis.
3. Name the phases of the cell cycle.
4. What are the three cell cycle checkpoints? What do they check?
5. How does binding of a growth factor induce the MAP kinase cascade?
6. What is the use of the MAP kinase cascade?
7. What is the use of programmed cell death?
8. What are the molecular components of apoptotic pathways?
9. What are the general components of a virus?
10. Why can viruses not replicate outside a host cell?

References

Berry, D. M., & Almeida, J. D. (1968). The morphological and biological effects of various antisera on avian infectious bronchitis virus. *The Journal of General Virology, 3*(1), 97–102.

Chang, H. C., Karim, F. D., O'Neill, E. M., Rebay, I., Solomon, N. M., Therrien, M., Wassarman, D. A., Wolff, T., & Rubin, G. M. (1994). Ras signal transduction pathway in Drosophila eye development. *Cold Spring Harbor Symposia on Quantitative Biology, 59*, 147–153.

Dick, F. A., & Rubin, S. M. (2013). Molecular mechanisms underlying RB protein function. *Nature Reviews. Molecular Cell Biology, 14*(5), 297–306.

Duesberg, P. H. (1987). Retroviruses as carcinogens and pathogens: Expectations and reality. *Cancer Research, 47*(5), 1199–1220.

Duesberg, P. H., & Robinson, W. S. (1966). Nucleic acid and proteins isolated from the Rauscher mouse leukemia virus (MLV). *Proceedings of the National Academy of Sciences of the United States of America, 55*(1), 219–227.

Echalier, A., Endicott, J. A., & Noble, M. E. (2010). Recent developments in cyclin-dependent kinase biochemical and structural studies. *Biochimica et Biophysica Acta, 1804*(3), 511–519.

Ellermann, V., & Bang, O. (1908). Experimentelle Leukemie bei Hühnern. *Centralblatt Bakteriologie und Parasitenkunde, 46*, 595–609.

Fung, T. S., & Liu, D. X. (2019). Human coronavirus: Host-pathogen interaction. *Annual Review of Microbiology, 73*, 529–557.

Georges, K. E. (1910). *Ausführliches lateinisch-deutsches Handwörterbuch.*

Hartwell, L. H., Culotti, J., Pringle, J. R., & Reid, B. J. (1974). Genetic control of the cell division cycle in yeast. *Science, 183*(4120), 46–51.

Hunter, T., & Eckhart, W. (2004). The discovery of tyrosine phosphorylation: It's all in the buffer! *Cell, 116*(2 Suppl), S35–S39, 31 p following S48.

Hunter, T., & Sefton, B. M. (1980). Transforming gene product of Rous sarcoma virus phosphorylates tyrosine. *Proceedings of the National Academy of Sciences of the United States of America, 77*(3), 1311–1315.

Kerr, J. F., Wyllie, A. H., & Currie, A. R. (1972). Apoptosis: A basic biological phenomenon with wide-ranging implications in tissue kinetics. *British Journal of Cancer, 26*(4), 239–257.

Koonin, E. V., & Yutin, N. (2018). Multiple evolutionary origins of giant viruses. *F1000Res, 7*, F1000 Faculty Rev-1840.

Lai, M. M., & Cavanagh, D. (1997). The molecular biology of coronaviruses. *Advances in Virus Research, 48*, 1–100.

Moelling, K., & Broecker, F. (2019). Viruses and evolution – Viruses first? A personal perspective. *Frontiers in Microbiology, 10*, 523.

Nurse, P., Thuriaux, P., & Nasmyth, K. (1976). Genetic control of the cell division cycle in the fission yeast Schizosaccharomyces pombe. *Molecular & General Genetics, 146*(2), 167–178.

Okada, Y. (1999). Historical overview of research on the tobacco mosaic virus genome: Genome organization, infectivity and gene manipulation. *Philosophical Transactions of the Royal Society of London. Series B, Biological Sciences, 354*(1383), 569–582.

Popgeorgiev, N., Temmam, S., Raoult, D., & Desnues, C. (2013). Describing the silent human virome with an emphasis on giant viruses. *Intervirology, 56*(6), 395–412.

Rous, P. (1967). The challenge to man of the neoplastic cell. *Cancer Research, 27*(11), 1919–1924.

Rozakis-Adcock, M., Fernley, R., Wade, J., Pawson, T., & Bowtell, D. (1993). The SH2 and SH3 domains of mammalian Grb2 couple the EGF receptor to the Ras activator mSos1. *Nature, 363*(6424), 83–85.

Rubin, H., & Temin, H. M. (1958). Infection with the Rous sarcoma virus in vitro. *Federation Proceedings, 17*(4), 994–1003.

Sefton, B. M., Hunter, T., Beemon, K., & Eckhart, W. (1980). Evidence that the phosphorylation of tyrosine is essential for cellular transformation by Rous sarcoma virus. *Cell, 20*(3), 807–816.

Sefton, B. M., Trowbridge, I. S., Cooper, J. A., & Scolnick, E. M. (1982). The transforming proteins of Rous sarcoma virus, Harvey sarcoma virus and Abelson virus contain tightly bound lipid. *Cell, 31*(2 Pt 1), 465–474.

Sender, R., & Milo, R. (2021). The distribution of cellular turnover in the human body. *Nature Medicine, 27*(1), 45–48.

Sender, R., Fuchs, S., & Milo, R. (2016). Revised estimates for the number of human and bacteria cells in the body. *PLoS Biology, 14*(8), e1002533.

Shi, Y. F., Szalay, M. G., Paskar, L., Sahai, B. M., Boyer, M., Singh, B., & Green, D. R. (1990). Activation-induced cell death in T cell hybridomas is due to apoptosis. Morphologic aspects and DNA fragmentation. *Journal of Immunology, 144*(9), 3326–3333.

Thomas, B. J., & Wassarman, D. A. (1999). A fly's eye view of biology. *Trends in Genetics, 15*(5), 184–190.

Index

A. Böttger, J. Moneer, *Basic Cell Biology for Dentists and Medical Students*, https://doi.org/10.1007/978-3-032-03636-0

N

O

P

W

Y

Y

Z

The manufacturer's authorised representative in the EU is Springer Nature Customer Service Centre GmbH, Europaplatz 3, 69115 Heidelberg, Germany. If you have any concerns regarding our products, please contact ProductSafety@springernature.com

Printed and bound by CPI Group (UK) Ltd, Croydon, CR0 4YY
07/07/2026
02160911-0005